Regulatory Status of Direct Food Additives

Editor:

Thomas E. Furia
President
Intechmark Corporation
Palo Alto, California

CRC Press
Taylor & Francis Group
Boca Raton London New York

CRC Press is an imprint of the
Taylor & Francis Group, an informa business

First published 1980 by CRC Press
Taylor & Francis Group
6000 Broken Sound Parkway NW, Suite 300
Boca Raton, FL 33487-2742

Reissued 2018 by CRC Press

Library of Congress Cataloging in Publication Data

Main entry under title:

Regulatory status of direct food additives.

Includes index.
1. Food additives--Law and legislation--United
States. 2. Food additives--Handbooks, manuals, etc.
I. Furia, Thomas E.
KF3879.P7R4 344.73'04232 80-11487
ISBN 0-8493-6380-2

A Library of Congress record exists under LC control number: 80011487

Publisher's Note
The publisher has gone to great lengths to ensure the quality of this reprint but points out that some imperfections in the original copies may be apparent.

Disclaimer
The publisher has made every effort to trace copyright holders and welcomes correspondence from those they have been unable to contact.

ISBN 13: 978-1-315-89720-2 (hbk)
ISBN 13: 978-1-351-07630-2 (ebk)

Visit the Taylor & Francis Web site at http://www.taylorandfrancis.com and the
CRC Press Web site at http://www.crcpress.com

PREFACE

The primary purpose of this volume is to provide readers of the *CRC Handbook of Food Additives* with quick summaries of and references to the current U.S. government and other regulations for some of the most frequently employed direct food additives. Indirect additives, such as pesticide residues, packaging adhesives, polymeric coatings, and animal feed chemicals, have not been included. Similarly, pending petitions, Food and Drug Administration (FDA) proposals, rejections of petitions, and petitions withdrawn without prejudice have not been included, since these represent, at best, a transient state or nonapproved use. Since regulations are far-reaching and changes are frequent, the ultimate source of reference is the Food and Drug Administration (FDA) and the official regulations appearing in the *Federal Register.*

The *Food Chemicals Codex,* prescribing minimum requirements of purity for chemicals intended for use in foods, has been endorsed by the Commissioner for the Food and Drug Administration as an outstanding guide and one that is regarded by the FDA as defining as "appropriate food grade." The editor concurs with this view, but he also stresses the absolute necessity for users of food chemicals to establish beyond question the quality of all ingredients to be employed in foods. In addition to becoming thoroughly familiar with the regulatory limits on use, the reader is urged to consult with additive suppliers to make certain that their products meet with approved U.S. government regulations regarding manufacture and purity. This aspect is mandatory for both suppliers and users of food additives, since all FDA clearances are based on principles of "good manufacturing practice."

The reader's attention is called to the following citations employed in Part II, and at the same time, he is cautioned to proper interpretations.

- **GMP** indicates *good manufactured practice.*
- **GRAS** indicates products *generally recognized as safe.* It is important to distinguish whether the product is sanctioned by a government agency, such as the FDA or MID, or by a trade association, such as FEMA. The FDA has the right to challenge trade association findings.
- **FEMA** signifies Flavor and Extract Manufacturers' Association. In the past the FDA has relied on the opinions of an expert committee of this organization to reach clearance decisions on flavor components. For many of the natural and synthetic flavors listed in Part II, the FEMA number, which indicates this association's GRAS status, has been included as an additional reference.
- **USDA** indicates the U.S. Department of Agriculture.
- **MID** refers to the Meat Inspection Division of the U.S. Department of Agriculture, which is responsible for clearing additives intended for use in all meats and meat products, with the exception of poultry. The USDA Poultry Division does not list approved additives, and it is recommended that the reader consult directly with this division concerning specific food substrates.
- **FR** signifies the *Federal Register,* the official U.S. government publication wherein all clearances, petition, withdrawals, time extensions, and other pertinent rulings are made public.
- **Food additive orders** are indicated by an FDA section number e.g., §121.2520, wherein the specific regulation is detailed. The latest version of such orders should be consulted, since limitations as to use, product specifications, and labeling are constantly changing.
- **Product specifications apply** indicate that only materials, manufactured by a specific process and/or having certain defined specifications, can be employed for the use described.

In the past, the Regulatory Status of Direct Food Additives was found in Part II of the *CRC Handbook of Food Additives* Volume I (1st and 2nd Editions). Because of the increased size of the information, we are now publishing this section as a separate volume and will update it as deemed necessary by developments in the field.

REGULATORY STATUS OF DIRECT FOOD ADDITIVES

Material	Alternate Name [Source]	FEMA No.	Regulation	Limitations
Abietic anhydride	See rosin [*Pinus palustris* Mill. and other *Pinus* spp.]			
Absinthium	See *artemisia* [*Artemisia absinthium* (L.) and other *Artemisia* spp.]			
Absinthium, extract	See wormwood, extract [*A. absinthium* (L.)]			
Absinthium, oil	See wormwood, oil [*A. absinthium* (L.)]			
Acacia	Gum acacia; gum arabic [*Acacia senegal* (L.) Willd., *A. verek* Guill. and Per.]	2001	§182.1	GRAS, stabilizer
Acetal	Acetaldehyde diethyl acetal	2002	FDA §172.515	Synthetic flavor/adjuvant
Acetaldehyde	Ethanal	2003	FDA §182.1	GRAS, synthetic flavor/adjuvant
Acetaldehyde benzyl β-methoxyethyl acetal	See benzyl methoxyethyl acetal			
Acetaldehyde butyl phenethylacetal		3125		
Acetaldehyde diethyl acetal	See acetal			
Acetaldehyde dimethyl acetal	See 1,1-dimethoxyethane			
Acetaldehyde phenethyl propyl acetal		2004	FDA §172.515	Synthetic flavor/adjuvant
Acetal R	See acetaldehyde phenethyl propyl acetal			
Acetanisol	See 1-(*p*-methoxyphenyl)-2-propanone			
Acetanisole	4′-Methoxyacetophenone	2005	FDA §172.515	Synthetic flavor/adjuvant
Acetate C-8	See octyl acetate			
Acetate C-9	See nonyl acetate			
Acetate C-10	See decyl acetate			
Acetate C-11	See 10-undecent-1-yl acetate			
Acetate C-12	See lauryl acetate			
Acetate PA	See allyl phenoxyacetate			

REGULATORY STATUS OF DIRECT FOOD ADDITIVES (continued)

Material	Alternate Name [Source]	FEMA No.	Regulation	Limitations
Acetic acid	Ethanoic acid	2006	FDA §182.1	GRAS, miscellaneous and/or general-purpose food additive; product specifications apply
			FDA §8.303	Production of caramel
			MID	To separate fatty acids and glycerol in rendered fats; refining agent must be eliminated during manufacturing process
Acetic acid, α-amylcinnamyl acetate	See α-amylcinnamyl acetate			
Acetic acid, benzyl ester	See benzyl acetate			
Acetic acid, perillyl ester	See perillyl acetate			
Acetic acid, phenoxy, allyl ester	See allyl phenoxyacetate			
Acetic acid, phenyl, allyl ester	See allyl phenylacetate			
Acetic acid, thioethyl ester	See ethyl thioacetate			
Acetic acid, thiopropyl ester	See propyl thioacetate			
Acetic aldehyde	See acetaldehyde			
Acetic anhydride	Acetic oxide		FDA §121.1030	Esterifier for food starch, in combination with adipic anhydride; adipic anhydride, not to exceed 0.12%; acetic anhydride, not to exceed 5.0%
Acetic ether	See ethyl acetate			
(tri-)Acetin	Glyceryl triacetrate; triacetin	2007	FDA §182.1	GRAS, miscellaneous and/or general-purpose food additive
Acetoacetic ester	See ethyl acetoacetate			
Acetoin	Acetyl methyl carbinol; 2,3-butanolone; dimethylketol; 3-hydroxy-2-butanone; γ-hydroxy-β-oxobutanone	2008	FDA §182.1	GRAS, synthetic flavor/adjuvant
Acetoin acetate	See acetyl methyl carbinyl acetate See 2-acetoxy-3 butanone			

2′-Acetonaphthone	See methyl β-naphthyl ketone			
Acetone	Dimethyl ketone	3326	FDA §121.1042	30 ppm in spice oleoresins, as a residue from extraction of spice
			FDA §8.390	Diluent in inks for marking fruits and vegetables, with no residue; product specifications apply
			FDA §8.305	Solvent for annatto extract; residue limit of 50 ppm
Acetone peroxides			FDA §121.1023	Manufacturing and bleaching of flour for bread and rolls where standards of identity do not preclude use; product specifications apply
Acetone propylene glycol acetal	See 2,2,4-trimethyl-1,3-oxacyclopentane			
Acetophenone	Methyl phenyl ketone	2009	FDA §172.515	Synthetic flavor/adjuvant
2-Acetoxy-3-butanone	See acetyl methyl carbinyl acetate			
	Acetoin acetate	3526		See *Food Technol.*, February 1978
9-Acetoxy-1-*p*-menthene	See 1-*p*-Menthen-9-yl acetate			
1-Acetoxy-2-methoxybenzene	See guaiacyl acetate			
2-Acetoxy-3-butanone	2-Butanon-3-yl acetate	3526		See *Food Technol.*, February 1978
4-Acetoxy-3-methoxy-1-(1-propen-1-yl) benzene	See isoeugenyl acetate			
9-Acetoxy-1-*p*-menthene	See 1-*p*-menthen-9-yl acetate			
3-Acetoxyoctene	See 1-octen-3-yl acetate			
Acetoyl butyrate	See butan-3-one-2-yl butanoate			
Acetylacetaldehyde, dimethyl acetal	See 3-oxobutanal, dimethyl acetal			
4-Acetylanisole	See acetanisole			
p-Acetylanisole	See acetanisole			
Acetylated monoglycerides			FDA §121.1018	Product specifications apply
			MID	Emulsifier for shortening
Acetylbenzene	See acetophenone			
Acetylbenzol	See acetophenone			
Acetyl benzoyl	See 1-phenyl-1,2-propanedione			
Acetyl butyryl	See 2,3-hexanedione			
Acetyl *o*-cresol	See *o*-tolyl acetate			
Acetyl *p*-cresol	See *p*-tolyl acetate			

REGULATORY STATUS OF DIRECT FOOD ADDITIVES (continued)

Material	Alternatate Name [Source]	FEMA No.	Regulation	Limitations
p-Acetylcumene	See *p*-isopropylacetophenone			
3-Acetyl-2,5-dimethylfuran		3391		
2-Acetyl-3,5(and 6)-dimethylpyrazine		3327		
3-Acetyl-2,5-dimethythiophene	2,5-Dimethyl-3-acetylthiophene	3527		From *Food Technol.*, February 1978
Acetyl ethyl carbinol	See 3-hydroxy-2-pentanone			
2-Acetyl-3-ethyl-1,4-diazine	See 2-acetyl-3-ethyl pyrazine			
2-Acetyl-3-ethyl pyrazine	2-Acetyl-3-ethyl-1,4-diazine	3250		See *Food Technol.*, May 1972
Acetyl eugenol	See eugenyl acetate			
Acetyl formaldehyde	See pyruvaldehyde			
Acetylformic acid	See pyruvic acid			
2-Acetylfuran	See 2-furyl methyl ketone			
Acetyl guaiacol	See guaiacyl acetate			
acetyl isobutyryl	See 4-methyl-2,3-pentanedione			
Acetyl isoeugenol	See isoeugenyl acetate			
Acetyl isopentanoyl	See 5-methyl-2,3-hexanedione			
Acetyl isovaleryl	See 5-methyl-2,3-hexanedione			
1-Acetyl-4-methylbenzene	See 4′-methylacetophenone			
Acetyl methyl carbinol	See acetoin			
Acetyl methyl carbinyl acetate		—		
β-Acetylnaphthalene	See methyl *β*-naphthyl ketone			
Acetyl nonanoyl	See 2,3-undecadione			
Acetyl nonyryl	See 2,3-undecadione			
Acetyl pelargonyl	See 2,3-undecadione			
Acetyl pentanoyl	See 2,3-heptanedione			
3-Acetyl propionic acid	See levulinic acid			
β-Acetyl propionic acid	See levulinic acid			
Acetyl propionyl	See 2,3-pentanedione			
Acetylpyrazine	Methyl pyrazinyl ketone	3126		10 ppm max in finished foods
2-Acetylpyridine		3251		See *Food Technol.*, May 1972

3-Acetylpyridine		3424		
β-Acetylpyridine	See 3-acetylpyridine			
2-Acetylpyrrole	See methyl-2-pyrrolyl ketone			
2-Acetylthiazole		3328		
p-Acetyltoluene	See 4′-methylacetophenone			
Acetyl tributyl citrate	See tributyl acetylcitrate			
Acetyl valeryl	See 2,3-heptanedione			
Acetyl vanillin	See vanillan acetate			
Acetyl-*m*-xylene	See 2,4-dimethylacetophenone			
Achilleic acid	See aconitic acid			
Acidophilus-type bacterial starters			MID	0.5% develops flavor in dry sausage, pork roll, thuringer, Lebanon bologna, cervelat, and salami; product specifications apply
Aconitic acid	Achilleic acid; citridic acid; equisetic acid; 2-carboxyglutaconic acid [*Aconitum napellus* (L.)]		FDA §182.1	GRAS, synthetic flavor/adjuvant
Acrolein	2-Propenal		FDA §121.1031	Etherification of food starch up to 0.6%; esterification and etherification of food starch up to 0.3%, with vinyl acetate up to 7.5%
Active valeric acid	See isovaleric acid			
Adipic acid	1,4-Butanedicarboxylic acid	2011	FDA §182.1 FDA §172.515	GRAS, buffer and neutralizing agent Synthetic flavor
Adipic anhydride			FDA §121.1031	Esterifier for food starch, in combination with acetic anhydride; adipic anhydride, not to exceed 0.12%, acetic anhydride, not to exceed 5.0%
Agar	Agar-agar [*Gelidium cartilagineum* (L.) Gaillon, *Gracilaria confervoides* (L.) Greville, and related red algae]	2012	FDA §182.1	GRAS, stabilizer
Agropyrum	See dog grass extract, [*Agropyron repens* (L.) Beauv.]			
Alanine (L- and DL-forms)			FDA §182.1	GRAS, nutrient and/or dietary supplement, as the free acid, hydrochloride salt, hydrate, or anhydrous form
β-Alanine		3252		

REGULATORY STATUS OF DIRECT FOOD ADDITIVES (continued)

Material	Alternate Name [Source]	FEMA No.	Regulation	Limitations
Alcannin, extract	See alkanet root, extract [*Alkanna tinctoria* Tausch.]			
Alcohol, anhydrous	See Ethyl alcohol			
Alcohol C-7	See heptyl alcohol			
Alcohol C-8	See 1-octanol			
Alcohol C-9	See nonyl alcohol			
Alcohol C-10	See 1-decanol			
Alcohol C-11, undecylic	See undecyl alcohol			
Alcohol C-12	See lauryl alcohol			
Alcohol C-16	See 1-hexadecanol			
Alcohol, denatured			FDA §8.390	Diluent in mixtures for coloring egg shells with no penetration of the mixture through egg shell into egg; product specifications apply
Alcohol, SDA-3A			FDA §8.390	Diluent in inks for marking fruits and vegetables, with no residue; product specifications apply
Alcohol, specially denatured			FDA §8.300	Diluent in inks for marking gum and confectionery and for marking food supplements in tablet form, with no residues; product specifications apply
Aldehyde C-6	See hexanal			
Aldehyde C-7	See heptanal			
Aldehyde C-8	See *n*-octanal			
Aldehyde C-8 dimethyl acetal	See octanal dimethyl acetal			
Aldehyde C-9	See *n*-nonanal			
Aldehyde C-10	See decanal			
Aldehyde C-10 dimethyl acetal	See decanal dimethyl acetal			
Aldehyde C-11 undecylenic	See 9-undecenal			
Aldehyde C-11 undecyclic	See undecanal			

Aldehyde C-12 lauric	See lauric aldehyde			
Aldehyde C-12 M.N.A.	See 2-methyl undecanal			
Aldehyde C-14 pure (so-called)	See γ-undecalactone			
Aldehyde C-14 myristic	See myristaldehyde			
Aldehyde C-16 pure (so-called)	See 3-methyl-3-phenyl glycidic acid ethyl ester			
Aldehyde C-18 (so-called)	See γ-nonalactone			
Alfalfa, extract	[*Medicago sativa* (L.)]	2013	FDA §182.1	Natural flavor
Alfalfa, herb and seed	[*M. sativa* (L.)]		FDA §182.1	GRAS, natural flavor and natural flavor extractive
Algae, brown	Kelp [*Laminaria* spp., *Nereocystis* spp.]		FDA §182.1	GRAS, natural substance and solvent-free natural extractive used with natural flavorings
Algae, red	[*Porphyra* spp., *Rhodymenia palmata* (L.) Grev.]		FDA §182.1	GRAS, natural substance and solvent-free natural extractive used with natural flavorings
Algin	[*Laminaria* spp. and other kelps]	2014	FDA §19	Up to 0.8% in specified cheeses; up to 0.5% by weight in specified cheeses
			FDA §20	In frozen desserts
			MID	Extender and stabilizer in breading mixes and sauces
Alginates — sodium, calcium, and ammonium salts		2015	FDA §182.1	GRAS, stabilizer
			FDA §172.515	Synthetic flavor/adjuvant
Alkaline aqueous solution			FDA §8.305	Extractant for annatto
Alkalis			FDA §8.305	Food-grade, to adjust pH; product specifications apply
Alkanet root, extract	[*Alkanna tinctoria* Tausch.]	2016	MID	Coloring sausage casings, oleomargarine, and shortening; marking or branding ink on products; product specifications apply
			FDA §172.510	
Alkannin, extract	See alkanet root, extract [*Alkanna tinctoria* Tausch.]			
Allerol	See nerol			
Allomaleic acid	See fumaric acid			

REGULATORY STATUS OF DIRECT FOOD ADDITIVES (continued)

Material	Alternate Name [Source]	FEMA No.	Regulation	Limitations
Allomalenic acid	See fumaric acid			
Allspice (see also pimenta)	[*Pimenta officinalis* Lindl.]	2017	FDA §182.1	GRAS, natural flavor and natural flavor extractive
Allspice, oil	[*P. officinalis* Lindl.]	2018	FDA §182.1	Natural flavor
Allspice, oleoresin	[*P. officinalis* Lindl.]		2 FDA §182.1	Natural flavor
Allura® Red AC	see FD&C Red No. 40			
Allyl acetic acid	See 4-pentenoic acid			
p-Allylanisole	See estragole			
Allyl anthranilate		2020	FDA §172.515	Synthetic flavor
Allyl butyrate		2021	FDA §172.515	Synthetic flavor/adjuvant
Allyl caproate	See allyl hexanoate			
Allyl caprylate	See allyl octanoate			
4-Allyl catechol-2-methyl ether	See eugenol			
Allyl cinnamate		2022	FDA §172.515	Synthetic flavor
Allyl crotonate		—		
Allyl cyclocitrylideneacetone	See allyl α-ionone			
Allyl cyclohexaneacetate		2023	FDA §172.515	Synthetic flavor
Allyl cyclohexanebutyrate		2024	FDA §172.515	Synthetic flavor
Allyl cyclohexanecaproate	See allyl cyclohexanehexanoate			
Allyl cyclohexanehexanoate		2025	FDA §172.515	Synthetic flavor
Allyl 6-cyclohexanehexanoate	See allyl cyclohexanevalerate			
Allyl cyclohexanepropionate	See allyl cyclohexanvalerate			
Allyl cyclohexanepropionate	Allyl 3-cyclohexylpropionate; allyl β-cyclohexylpropionate	2026	FDA §172.515	Synthetic flavor
Allyl cyclohexanevalerate		2027	FDA §17.515	Synthetic flavor
Allyl cyclohexylacetate	See allyl cyclohexaneacetate			
Allyl 4-cyclohexylbutyrate	See allyl cyclohexanebutyrate			
Allyl cyclohexyl-*n*-butyrate	See allyl cyclohexanebutyrate			

Allyl 5-cyclohexylpentanoate	See allyl cyclohexanevalerate			
Allyl 3-cyclohexylpropionate	See allyl cyclohexanepropionate			
Allyl β-cyclohexylpropionate	See allyl cyclohexanepropionate			
Allyl cyclohexylvalerate	See allyl cyclohexanevalerate			
4-Allyl-1,2-dimethoxybenzene	See eugenyl methyl ether			
Allyl disulfide	Diallyl disulfide	2028	FDA §172.515	Synthetic flavor
Allyl enanthate	See allyl heptanoate			
Allyl 2-ethylbutyrate		2029	FDA §172.515	Synthetic flavor
Allyl 2-furoate		2030		
4-Allylguaiacol	See eugenol			
Allyl heptanoate		2031		
Allyl heptoate	See allyl heptanoate			
Allyl heptylate	See allyl heptanoate			
Allyl 2,4-hexadienoate	See allyl sorbate			
Allyl hexahydrophenylacetate	See allyl cyclohexaneacetate			
Allyl hexahydrophenylbutyrate	See allyl cyclohexanebutyrate			
Allyl hexahydrophenylhexanoate	See allyl cyclohexanehexanoate			
Allyl hexahydrophenylpropionate	See allyl cyclohexanepropionate			
Allyl hexahydrophenylvalerate	See allyl cyclohexanevalerate			
Allyl hexanoate	Allyl caproate	2032	FDA §172.515	Synthetic flavor/adjuvant
Allyl hexenoate		—		
Allyl 2-hexenoate	See allyl hexenoate			
α-Allylionone	See allyl α-ionone			
Allyl α-ionone	1-(2,6,6-Trimethyl-2-cyclohexen-1-yl)-1,6-heptadien-3-one	2033	FDA §172.515	Synthetic flavor
Allyl isopentanoate	See allyl isovalerate			
Allyl isothiocyanate	Mustard oil	2034	FDA §172.515	Synthetic flavor
Allyl isovalerate		2045	FDA §172.515	Synthetic flavor
Allyl isovalerianate	See allyl isovalerate			

REGULATORY STATUS OF DIRECT FOOD ADDITIVES (continued)

Material	Alternate Name [Source]	FEMA No.	Regulation	Limitations
Allyl mercaptan	2-Propene-1-thiol	2035	FDA §172.515	Synthetic flavor
4-Allyl-2-methoxyphenol	See eugenol			
4-Allyl-2-methoxyphenyl acetate	See eugenyl acetate			
4-Allyl-2-methoxyphenyl benzoate	See eugenyl benzoate			
4-Allyl-2-methoxyphenyl formate	See eugenyl formate			
Allyl β-methylacrylate	See allyl crotonate			
Allyl 3-methylbutanoate	See allyl isovalerate			
Allyl *trans*-2-methyl-2-butenoate	See allyl tiglate			
Allyl methyl disulfide		3127		
Allyl methyl trisulfide		3253		
Allyl nonanoate		2036	FDA §172.515	Synthetic flavor
Allyl nonylate	See allyl nonanoate			
Allyl octanoate		2037	FDA §172.515	Synthetic flavor
Allyl octylate	See allyl octanoate			
Allyl pelargonate	See allyl nonanoate			
Allyl phenoxyacetate	Acetate PA	2038	FDA §172.515	Synthetic flavor
Allylphenylacetate		2039	FDA §172.515	Synthetic flavor
Allyl β-phenylacrylate	See allyl cinnamate			
Allyl 3-phenyl-2-propenoate	See allyl cinnamate			
Allyl propanoate	See allyl propionate			
Allyl propionate		2040	FDA §172.515	Synthetic flavor
Allyl sorbate	Allyl 2,4-hexadienoate	2041	FDA §172.515	Synthetic flavor
Allyl sulfhydrate	See allyl mercaptan			
Allyl sulfide	Thioallyl ether; diallyl sulfide	2042	FDA §172.515	Synthetic flavor
Allythiol	See allyl mercaptan			
Allyl thiopropionate		3329		
Allyl tiglate	Allyl *trans*-2-methyl-2-butenoate	2043	FDA §172.515	Synthetic flavor

Allyl α-toluate	See allyl phenylacetate			
Allyl trisulfide	See diallyl trisulfide			
Allyl 10-undecenoate	Allyl undecylenate	2044	FDA §172.515	Synthetic flavor
Allyl undecylenate	See allyl 10-undecenoate			
4-Allylveratrole	See eugenyl methyl ether			
Almond, bitter, oil (ffpa)	[*Prunus amygdalus* Batsch; also from *P. armeniaca* (L.) or *P. persica* (L.) Batsch]	2046	FDA §182.1	GRAS, natural flavor extractive when free from prussic acid
Aloe, extract	[*Aloe perryi* Baker; *A. barbadensis* Mill.; *A. ferox* Mill.; and hybrids of this species with *A. africana* Mill. and *A. spicata* Baker]	2047	FDA §172.510	Natural flavor
Althea, flowers	[*Althaea officinalis* (L.)]	—	FDA §172.510	
Althea, root	[*Althaea officinalis* (L.)]	2048	FDA §172.510	Natural flavor
Alumina			FDA §9.100	For manufacture of FD&C color lakes; product specifications apply
Aluminum ammonium sulfate			FDA §182.1	GRAS, miscellaneous and/or general-purpose food additive
Aluminum calcium silicate			FDA §22.8 and §22.9	In vanilla powder and vanilla-vanillin powder under food standards
			FDA §182.1	GRAS, anticaking agent in table salt up to 2%
Aluminum nicotinate			FDA §121.1141	Source of niacin in foods for special dietary use; product specifications apply
Aluminum oxide			FDA §8.316 and §8.6005	Dispersing agent in mixtures for food or drug use containing titanium dioxide, up to 2%
Aluminum potassium sulfate			FDA §182.1	GRAS, miscellaneous and/or general-purpose food additive
Aluminum salts of fatty acids			FDA §121.1071	Binder, emulsifier, and anticaking agent in food; product specifications apply
Aluminum sodium sulfate			FDA §182.1	GRAS, miscellaneous and/or general-purpose food additive
Aluminum stearate			FDA §121.1071 and §121.1099	Component of defoaming agents limited to use in processing beet sugar and yeast
Aluminum sulfate			FDA §182.1	GRAS, miscellaneous and/or general-purpose food additive

REGULATORY STATUS OF DIRECT FOOD ADDITIVES (continued)

Material	Alternate Name [Source]	FEMA No.	Regulations	Limitations
			FDA §121.1031	Modifier for food starch, up to 2.0% in combination with not more than 2.0% of 1-octenyl succinic anhydride
Ambergris	[*Physeter macrocephalus* (L.)]		FDA §182.1	GRAS, natural extractive
Ambergris, tincture		2049		
Ambreine, oil	See labdanum, oil [*Cistus* spp.]			
Ambrette, absolute, oil	[*Hibiscus abelmoschus* (L.)]	2050		
Ambrette, seed	[*H. moschatus* Moench; *H. abelmoschus* (L.)]		FDA §182.1	GRAS, natural flavor
Ambrette, seed, oil	[*H. abelmoschus* (L.)]	2051	FDA §182.1	Natural flavor
Ambrette, tincture	[*H. abelmoschus* (L.)]	2052		
Ambrettolide	See ω-6-hexadecenlactone			
Aminoacetic acid	See glycine			
1-Aminobutane	See butylamine			
DL-(3-Amino-3-carboxypropyl) dimethylsulfonium chloride		3445		
Aminoethanoic acid	See glycine			
2-Aminoethylbenzene	See phenethylamine			
Aminoglutaric acid	See glutamic acid			
α-Aminoisocaproic acid	See *l*-leucine			
1-Aminoisopentane	See isopentylamine			
α-Aminoisovaleric	See *d,l*-valine			
2-Amino-3-mercaptopropanoic acid	See *l*-cysteine			
α-Amino-β-mercaptopropanoic acid	See *l*-cysteine			
α-Amino-β-mercaptopropionic acid	See *l*-cysteine			
o-Aminomethylbenzoate	See methyl anthranilate			
2-Amino-3-methylbutanoic acid	See *d,l*-valine			

2-Amino-3-methylbutyric acid	See *d,l*-valine			
2-Amino-3-methylpentanoic acid	See *d,l*-isoleucine			
2-Amino-4-methylpentanoic acid	See *l*-leucine			
2-Amino-4-(methylthio)-butanoic acid	See *d,l*-methionine			
α-Amino-γ-(methylthio)-butyric acid	See *d,l*-methionine			
2-Amino-4-methylvaleric acid	See *l*-leucine			
α-Amino-β-methylvaleric acid	See *d,l*-isoleucine			
1-Amino-2-phenylethane	See phenethylamine			
2-Aminopropanoic acid	See β-alanine			
β-Aminopropionic acid	See β-alanine			
Ammonium alginate	See alginates — sodium, calcium, and ammonium salts			
Ammonium bicarbonate			FDA §182.1	GRAS, miscellaneous and/or general-purpose food additive
Ammonium carbonate			FDA §182.1	GRAS, miscellaneous and/or general-purpose food additive
			FDA §8.303	Including bicarbonate in production of caramel
Ammonium caseinate			FDA §20	*In frozen desserts*
Ammonium chloride			FDA §17	In bakery products; alone or with calcium sulfate and/or dicalcium phosphate, up to 0.25% by weight of flour
Ammonium hydroxide			FDA §182.1	GRAS, miscellaneous and/or general-purpose food additive
			FDA §8.303	In production of caramel
Ammonium isovalerate		2054	FDA §172.515	Synthetic flavor
Ammonium isovalerianate	See ammonium isovalerate			
Ammonium persulfate			FDA §121.1031	Bleaching agent for food starch, up to 0.075%; with sulfur dioxide, up to 0.05%

REGULATORY STATUS OF DIRECT FOOD ADDITIVES (continued)

Material	Alternate Name [Source]	FEMA No.	Regulation	Limitations
Ammonium phosphate (mono- and dibasic)			FDA §182.1	GRAS, miscellaneous and/or general-purpose food additive
			FDA §8.303	In production of caramel
Ammonium saccharin			FDA §182.1	GRAS, nonnutritive sweetener
Ammonium sulfate			FDA §182.1	GRAS, miscellaneous and/or general-purpose food additive
			FDA §8.303	In production of caramel
Ammonium sulfide		2053	FDA §172.515	Synthetic flavor
Ammonium sulfite			FDA §8.303	In production of caramel
Amyl (iso) acetate	See isoamyl acetate			
Amyl acetate, common	See isoamyl acetate			
iso-Amyl acetoacetate	See isoamyl acetoacetate			
Amyl (iso) acetoacetate	See isoamyl acetoacetate			
Amyl alcohol	Pentyl alcohol	2056	FDA §172.515	Synthetic flavor
Amyl (iso) alcohol	See isoamyl alcohol			
sec- n-Amyl alcohol	See 2-pentanol			
Amyl alcohol, commercial	See fusel, oil, refined			
Amyl aldehyde	See valeraldehyde			
Amyl (iso) aldehyde	See 3-methylbutyraldehyde			
Amyl (iso) amine	See isopentylamine			
Amyl (iso) benzoate	See isoamyl benzoate			
Amyl butanoate	See amyl butyrate			
Amyl (iso) butanoate	See isoamyl butyrate			
Amyl butyrate	Pentyl butyrate	2059	FDA §172.515	Synthetic flavor
Amyl (iso) butyrate	See isoamyl butyrate			
Amyl ethyl carbinyl acetate	See 3-octyl acetae			
iso-Amyl isobutyrate	See 3-methylbutyl 2-methyl propanoate			
γ-Amyl butyrolactone	See γ-nonalactone			
Amyl caprate	See amyl decanoate			
Amyl caprinate	See amyl decanoate			

Amyl caproate	See amyl hexanoate			
Amyl (iso) caproate	See isoamyl hexanoate			
Amyl caprylate	See amyl octanoate			
Amyl (iso) caprylate	See isoamyl octanoate			
Amyl cinnamal	See α-amylcinnamaldehyde			
α-*n*-Amylcinnamal dimethyl acetal	See α-amylcinnamaldehyde dimethyl acetal			
α-Amylcinnamaldehyde	α-Pentylcinnamaldehyde	2061	FDA §172.515	Synthetic flavor
α-Amylcinnamaldehyde dimethyl acetal	1,1-Dimethoxy-2-amyl-3-phenyl-2-propene	2062	FDA §172.515	Synthetic flavor
Amyl (iso) cinnamate	See isoamyl cinnamate			
n-Amyl cinnamic alcohol	See α-amylcinnamyl alcohol			
Amylcinnamic aldehyde	See α-amylcinnamaldehyde			
Amylcinnamyl acetate	See α-amylcinnamyl acetate			
α-Amylcinnamyl acetate	α-Pentylcinnamyl acetate	2064	FDA §172.515	Synthetic flavor
α-Amylcinnamyl alcohol	α-Pentylcinnamyl alcohol	2065	FDA §172.515	Synthetic flavor
α-Amylcinnamyl formate	α-Pentylcinnamyl formate	2066	FDA §172.515	Synthetic flavor
α-Amylcinnamyl isovalerate	α-Pentylcinnamyl isovalerate	2067	FDA §172.515	Synthetic flavor
α-Amylcinnamyl isovalerianate	See α-amylcinnamyl isovalerate			
Amyl crotonyl acetate	See 1-octen-3-yl acetate			
Amyl decanoate		—		
2-Amyl-5 or 6-keto-1,4-dioxane		2076		
Amyl (iso) dodecanoate	See isoamyl laurate			
Amyl ethyl carbinol	See 3-octanol			
Amyl ethyl ketone	See 3-octanone			
Amyl formate	Pentyl formate	2068	FDA §172.515	Synthetic flavor
Amyl (iso) formate	See isoamyl formate			
Amyl formiate	See amyl formate			
Amyl (iso) 2-furanbutyrate	See isoamyl 2-furanbutyrate			
Amyl (iso) 2-furanpropionate	See isoamyl 2-furanpropionate			
α-Amyl (iso) furfurylacetate	See isoamyl 2-furanbutyrate			
Amyl 2-furoate	Pentyl 2-furoate	2072		
Amyl heptanoate	Pentyl heptanoate	2073	FDA §172.515	Synthetic flavor

REGULATORY STATUS OF DIRECT FOOD ADDITIVES (continued)

Material	Alternatate Name [Source]	FEMA No.	Regulation	Limitations
Amyl heptoate	See amyl heptanoate			
Amyl heptylate	See amyl heptanoate			
Amyl hexanoate	Pentyl hexanoate	2074	FDA §172.515	Synthetic flavor
Amyl (iso) hexanoate	See isoamyl hexanoate			
Amyl 2-hydroxybenzoate	See amyl salicylate			
Amyl (iso) *o*-hydroxybenzoate	See isoamyl salicylate			
4-*n*-Amyl-4-hydroxybutyric acid lactone	See γ-nonalactone			
5-*n*-Amyl-5-hydroxypentanoic acid lactone	See δ-decalactone			
Amyl (2-hydroxyphenyl) formate	See amyl salicylate			
Amyl (iso) isobutyrate	See isoamyl isobutyrate			
Amyl (iso) isopentanoate	See isoamyl isovalerate			
Amyl isovalerate		—		
Amyl (iso) isovalerate	See isoamyl isovalerate			
Amyl (iso) β-ketobutyrate	See isoamyl acetoacetate			
Amyl (iso) laurate	See isoamyl laurate			
Amyl methanoate	See amyl formate			
Amyl (iso) methanoate	See isoamyl formate			
Amyl 3-methylbutanoate	See amyl isovalerate			
Amyl (iso) 2-methylbutanoate	See isoamyl 2-methylbutyrate			
Amyl (iso) 3-methylbutanoate	See isoamyl isovalerate			
Amyl methyl carbinol	See 2-heptanol			
Amyl methyl ketone	See 2-heptanone			
Amyl (iso) 2-methyl propanoate	See isoamyl isobutyrate			
iso-Amyl 2-methylpropanoate	See 3-methylbutyl 2-methylpropanoate			

Amyl (iso) nonanoate	See isoamyl nonanoate			
Amyl (iso) nonylate	See iso amyl nonanoate			
Amyl octanoate	Pentyl octanoate	2079	FDA §172.515	Synthetic flavor
Amyl (iso) octanoate	See isoamyl octanoate			
Amyl octylate	See amyl octanoate			
Amyl (iso) octylate	See isoamyl octanoate			
Amyl (iso) oxobutanoate	See isoamyl acetoacetate			
Amyl (iso) pelargonate	See isoamyl nonanoate			
Amyl phenylacetate		—		
Amyl (iso) phenylacetate	See isoamyl phenylacetate			
α-Amyl-β-phenylacrolein	See α-amylcinnamaldehyde			
α-*n*-Amyl-β-phenylacrolein dimethyl acetal	See α-amylcinnamaldehyde dimethyl acetal			
α-*n*-Amyl-β-phenylacryl acetate	See α-amylcinnamyl acetate			
Amyl (iso) β-phenylacrylate	See isoamyl cinnamate			
α-*n*-Amyl-β-phenylacryl formate	See α-amylcinnamyl formate			
α-*n*-Amyl-β-phenylacryl isovalerate	See α-amylcinnamyl isovalerate			
α-*n*-Amyl-β-phenylacryl 3-methylbutanoate	See α-amylcinnamyl isovalerate			
2-Amyl-3-phenyl-2-propen-1-ol	See α-amylcinnamyl alcohol			
Amyl (iso) propanoate	See isoamyl propionate			
Amyl (iso) propionate	See isoamyl propionate			
Amyl (iso) pyruvate	See isoamyl pyruvate			
Amyl salicylate		—		
Amyl (iso) salicylate	See isoamyl salicylate			
Amyl (iso) α-toluate	See isoamyl phenylacetate			
δ-*n*-Amyl-δ-valerolactone	See δ-decalactone			
Amyl vinyl carbinol	See 1-octen-3-ol			
Amyl vinyl carbinol acetate	See 1-octen-3-yl acetate			
Amyl vinyl carbinyl acetate	See 1-octen-yl acetate			
Amyl vinyl ketone	See 1-octen-3-one			
Amyris	West Indian sandalwood [*Amyris balsamifer* (L.)]	—	FDA §172.510	Natural flavor
Amyris, oil	[*A. balsamifera* (L.)]	—	FDA §172.510	

REGULATORY STATUS OF DIRECT FOOD ADDITIVES (continued)

Material	Alternate Name [Source]	FEMA No.	Regulation	Limitations
Anchusin, extract	See alkanet root, extract [*Alkanna tinctoria* Tausch.]			
Anethole	*p*-Propenyl anisole	2086	FDA §182.1	GRAS, synthetic flavor/adjuvant
Angelica lactone	See ω-pentadecalactone			
α-Angelica lactone	See 4-hydroxy-3-pentenoic acid lactone			
Angelica root, extract	[*Angelica archangelica* (L.) or other species of *Angelica*]	2087	FDA §182.1	GRAS, natural flavor extractive
Angelica root, oil	[*A. archangelica* (L.)]	2088	FDA §182.1	GRAS, natural flavor extractive
Angelica seed, extract	[*A. archangelica* (L.)]	2089	FDA §182.1	Natural flavor
Angelica seed, oil	[*A. archangelica* (L.)]	2090	FDA §182.1	Natural flavor
Angelica stem, oil	[*A. archangelica* (L.)]	2091	FDA §182.1	
Angola weed	[*Roccella fuciformis* Arch.]	—	FDA §172.510	Natural flavoring use in alcoholic beverages only
Angostura®, extract	Cusparia [*Galipea officinalis* Hancock (*G. cusparia* D.C.)]	2092	FDA §182.1	GRAS, natural flavor and natural flavor extractive
Anisalcohol	See anisyl alcohol			
Anisaldehyde	See *p*-methoxybenzaldehyde			
Anise	[*Pimpinella anisum* (L.)]	2093	FDA §182.1	GRAS, natural flavor and natural flavor extractive
Anise, alcohol	See anisyl alcohol			
Anise, camphor	See anethole			
Anise, oil	[*P. anisum* (L.)]	2094	FDA §182.1	
Anise, star	[*Illicium verum* Hook. f.]	2095	FDA §182.1	GRAS, natural flavor
Anise, star, oil	[*I. verum* Hook. f.]	2096		
Anisic alcohol	See anisyl alcohol			
Anisic aldehyde	See *p*-methoxybenzaldehyde			
Anisic ketone	See 1-(*p*-methoxyphenyl)-2-propanone			
Anisole	Methoxybenzene	2097	FDA §172.515	Synthetic flavor
Anisyl acetate	*p*-Methoxybenzyl acetate	2098	FDA §172.515	Synthetic flavor
Anisyl acetone	See (4-(*p*-methoxyphenyl)-2-butanone			
p-Anisyl acetone	See 1-(*p*-methoxyphenyl)-2-propanone			
Anisyl alcohol	*p*-Methoxybenzyl alcohol	2099	FDA §172.515	Synthetic flavor

Anisyl alcohol, acetate	See anisyl acetate			
Anisyl alcohol, formate	See anisyl formate			
Anisyl butyrate	*p*-Methoxybenzyl butyrate	2100	FDA §172.515	Synthetic flavor
Anisyl formate	*p*-Methoxybenzyl formate	2101	FDA §172.515	Synthetic flavor
Anisyl methanoate	See anisyl formate			
Anisyl methyl ketone	See acetanisole			
Anisyl methyl ketone	See 1-(*p*-methoxyphenyl)-2-propanone			
Anisyl phenylacetate		—	FDA §172.515	Synthetic flavor
Anisyl propionate	*p*-Methoxybenzyl propionate	2102	FDA §172.515	Synthetic flavor
Anisyl α-toluate	See anisyl phenylacetate			
Annatto	[*Bixa orellana* (L.)]		MID	Coloring sausage casings, oleomargarine, or shortening; marking or branding inks on products; product specifications apply
Annatto, extract	[*B. orellana* (L.)]	2103	FDA §8.6003	Coloring ingested drugs generally; conforming with §8.305; product specifications apply
			FDA §8.305	Coloring foods generally; product specifications apply
Annatto, seed	[*B. orellana* (L.)]	2104	FEMA-GRAS	
Anotta, extract	See annatto, extract [*B. orellana* (L.)]			
Anotta, seed	See annatto, seed [*B. orellana* (L.)]			
β-Apo-8′-carotenal			FDA §8.302	Food color; not to exceed 16 mg/lb of solid or semisolid food or 15 mg/pt of liquid food; product specifications apply
Apple acid	See 1-malic acid			
Apricot kernel, oil	Persic oil; bitter almond [*Prunus armeniaca* (L.)]	2105	FDA §182.1	GRAS, *solvent-free* natural extractive used with natural flavorings
Arabic gum	See acacia, gum [*Acacia senegal* (L.) Willd.]			
Arabinogalactan	Western larch wood	3254	FDA §121.1174	Emulsifier, stabilizer, binder, or bodying agent in essential oils, nonnutritive sweeteners, flavor bases, nonstandardized dressings, and pudding mixes; product specifications apply; see *Food Technol.*, May 1972, for FEMA average max use level

REGULATORY STATUS OF DIRECT FOOD ADDITIVES (continued)

Material	Alternate Name [Source]	FEMA No.	Regulation	Limitations
L-Arabinose		3255		See *Food Technol.*, May 1972
Araboascorbic acid	See erythrobic acid			
Arborvitae	See cedar, white			
Argeol	See santalol (α and β)			
Arginine (L- and DL-forms)			FDA §182.1	GRAS, nutrient and/or dietary supplement as the free acid, hydrochloride salt, hydrate, or anhydrous form
Arnica, flowers	[*Arnica montana* (L.); *A. fulgens* Pursh.; *A. sororia* Greene; *A. cordifolia* Hooker]	—	FDA §172.510	Natural flavor
Arnotta, extract	See annatto, extract [*Bixa orellana* (L.)]			
Arnotta, seed	See annatto, seed [*B. orellana* (L.)]			
Artemisia	Wormwood absinthium [*Artemisia absinthium* (L.) and other *Artemisia* spp.]	3114	FDA §172.510	Natural flavor; thujone-free
Artichoke, leaves	[*Cynara scolymus* (L.)]	—	FDA §172.510	Natural flavoring used in alcoholic beverages only
Artifical smoke flavoring			MID	To flavor various meat products
Asafetida, fluid extract	[*Ferula assafoetida* (L.) and related *Ferula* spp.]	2106	FDA §182.1	GRAS, natural flavor extractive
Asafetida, gum	[*F. assafoetida* (L.)]	2107	FDA §182.1	Natural flavor
Asafetida, oil	[*F. assafoetida* (L.)]	2108	FDA §182.1	Natural flavor
Ascorbic acid	Vitamin C	2109	FDA §182.1	GRAS, nutrient and/or dietary supplement, chemical preservative
			MID	To accelerate color fixing in cured pork and beef cuts and cured comminuted meat food products; product specifications apply
iso-Ascorbic acid	See erythrobic acid			

Ascorbyl palmitate			FDA §182.1	GRAS, chemical preservative
Ash bark, prickly extract	[*Xanthoxylum* spp.]	2110	FDA §182.1	Natural flavor
Asparagus, seed and root, extract		—		
Aspartic acid (L- and DL-forms)			FDA §182.1	GRAS, nutrient and/or dietary supplement as the free acid, hydrochloride salt, hydrate, or anhydrous form
Aspergillus flavus-oryzae group			MID	With water, salt, and monosodium glutamate to soften tissues of beef cuts; tissue weight gain, 3% max
Aspergillus oryzae			FDA §17.1	In bakery products
			MID	With water, salt, and monosodium glutamate to soften tissues in beef cuts; tissue weight gain, 3% max
Assafoetida	See asafetida			
Attar of roses	See rose, Bulgarian, true Otto, oil [*Rosa damascena* Mill.]			
Aubepine liquid	See *p*-methoxybenzaldehyde			
Autolyzed yeast extract	See yeast extract, autolyzed			
Azodicarbonamide			FDA §121.1085	Aging and bleaching ingredient in cereal flour, up to 45 ppm; product specifications apply
Azole	See pyrrole			
Bacterial catalase	[*Micrococcus lysodeikticus*]		FDA §19.500	In cheddar cheese manufacture for removal of hydrogen peroxide; not to exceed 20 ppm of weight of milk treated
			FDA §121.1170	Destroying and removing hydrogen peroxide used in manufacture of cheese; product specifications apply
Bacterial starters			MID	Harmless bacterial starters of the acidophilus type, lactic acid starters, or cultures of *Pediococcus cerevisiae*, to develop flavor in dry sausage, pork roll, thuringer, Lebanon bologna, cervelat, and salami, at 0.5%
Baking powder			FDA	GRAS
Balm	Lemon balm; melissa [*Melissa officinalis* (L.)]	2111	FDA §182.1	GRAS, natural flavor and natural flavor extractive

REGULATORY STATUS OF DIRECT FOOD ADDITIVES (continued)

Material	Alternate Name [Source]	FEMA No.	Regulation	Limitations
Balm, leaves, extract	[*M. officinalis* (L.)]	2112	FDA §182.1	Natural flavor
Balm, lemon	See balm [*M. officinalis* (L.)]			
Balm, lemon, extract	See balm, leaves, extract [*M. officinalis* (L.)]			
Balm, lemon, oil	See balm, oil [*M. officinalis* (L.)]			
Balm, oil	[*M. officinalis* (L.)]	2113	FDA §182.1	Natural flavor
Balsam copáiba	See copaiba [South American spp. of *Copaifera* (L.)]			
Balsam fir, needles and twigs	Balsam fir [*Abies balsamea* (L.) Mill.]		FDA §172.510	Natural flavor
Balsam fir, oil	[*A. balsamea* (L.) Mill.]	2114		
Balsam fir, oleoresin	[*A. balsamea* (L.) Mill.]	2115		
Balsam of Peru	See Peruvian balsam			
Baros	See borneol			
Basil	[*Ocimum basilicum* (L.)]	2118	FDA §182.1	GRAS, natural flavor and natural extractive
Basil, bush	Dwarf basil [*O. minimum* (L.)]		FDA §182.1	GRAS, natural flavor
Basil, oil	[*Ocimum basilicum* (L.)]	2119	FDA §182.1	Natural flavor
Basil, oleoresin	[*O. basilicum* (L.)]	2120	FDA §182.1	Natural flavor
Bay leaf, West Indian, extract	[*Pimenta acris* Kostel]	2121	FDA §182.1	Natural flavor
Bay leaf, West Indian, oil	Myrcia oil [*P. racemosa* (Mill.) J. W. Moore; *P. acris* Kostel.]	2122	FDA §182.1	GRAS, natural flavor extractive
Bay leaf, West Indian, oleoresin	[*P. acris* Kostel.]	2123	FDA §182.1	Natural flavor
Bay, sweet	Bay; laurel [*Laurus nobilis* (L.)]	2124	FDA §182.1	GRAS, natural flavor
Bay, sweet, oil	[*L. nobilis* (L.)]	2125	FDA §182.1	GRAS, natural flavor extractive
BDS	See benzyl disulfide			
Beechwood creosote			FDA §172.515	Synthetic flavor
Beeswax, white	Bleached yellow wax; cire d'abeille absolute [*Apis mellifera* (L.)]	2126	FDA §172.510	Natural flavor

			FDA §182.1	GRAS, miscellaneous and/or general-purpose food additive
Beeswax, yellow	Yellow wax		FDA §182.1	GRAS, miscellaneous and/or general-purpose food additive
Beets, dehydrated	Beet powder		FDA §8.321	Coloring foods generally; product specifications apply
Benihinal	See 2-formyl-6,-dimethylbicyclo(3.1.1)hept-2-ene			
Bentonite	Wilkinite		FDA §182.1	GRAS, miscellaneous and/or general-purpose food additive
Benzaldehyde	Benzoic aldehyde; benzenecarbonal	2127	FDA §182.1	GRAS, synthetic flavor/adjuvant
Benzaldehyde, cyclic acetal with glycerol	See benzaldehyde glyceryl acetal			
Benzaldehyde dimethyl acetal		2128	FDA §172.515	Synthetic flavor
Benzaldehyde glyceryl acetal	2-Phenyl-*m*-dioxan-5-ol	2129	FDA §172.515	Synthetic flavor
Benzaldehyde propylene glycol acetal	4-Methyl-2-phenyl-*m*-dioxolane	2130	FDA §172.515	Synthetic flavor
Benzalglycerin	See benzaldehyde glyceryl acetal			
1-Benzazine	See quinoline			
Benzenecarbonal	See benzaldehyde			
Benzene carboxaldehyde	See benzaldehyde			
Benzene carboxylic acid	See benzoic acid			
1,3-Benzenediol	See resorcinol			
Benzene formic acid	See benzoic acid			
Benzenemethanethiol	See benzyl mercaptan			
Benzene, methoxy	See anisole			
Benzene, 1-methoxy-4-(1-propenol)	See anethole			
Benzenemethylal	See benzaldehyde			
Benzenepropenoic acid	See cinnamic acid			
Benzenethiol	Thiophenol	—	FDA §172.515	Synthetic flavor
Benzilideneacetone	See 4-phenyl-3-buten-2-one			
1,2-Benzisothiazolin-3-one,1,1-dioxide, sodium salt	See saccharin, sodium salt			
1,2-Benzodihydropyrone	See dihydrocoumarin			

REGULATORY STATUS OF DIRECT FOOD ADDITIVES (continued)

Material	Alternate Name [Source]	FEMA No.	Regulations	Limitations
Benzoe	See benzoin, resin [*Styrax* spp.]			
2-Benzofurancarboxaldehyde	2-Formylbenzofuran	3128		
Benzoic acid	Benzene carboxylic acid; dracylic acid; phenylformic acid	2131	FDA §182.1	GRAS, chemical preservative, up to 0.1%
			MID	To retard flavor reversion in oleomargarine, at 0.1%
Benzoic acid, benzyl ester	See benzyl bensoate			
Benzoic acid, 4-hydroxy, butyl ester	See butyl *p*-hydroxybenzoate			
Benzoic acid, 2-hydroxy, phenylmethyl ester	See benzyl salicylate			
Benzoic acid, isopentyl ester	See isoamyl benzoate			
Benzoic aldehyde	See benzaldehyde			
Benzoin	2-Hydroxy-2-phenyl-acetophenone	2132	FDA §172.515	Synthetic flavor
			FDA §8.390	Diluent in inks for marking fruits and vegetables; product specifications apply
(tri-)Benzoin	See glyceryl tribenzoate			
Benzoin resin	[*Styrax benzoin* Dryander; *S. paralleloneurus* Perkins; *S. tonkinensis* (Pierre) Craib; or other species of the section Anthostyrax of the genus *Styrax*]	2133	FDA §172.510	Natural flavor
Benzophenone	Diphenyl ketone	2134	FDA §172.515	Synthetic flavor
Benzo(6)pyridine	See quinoline			
1-Benzo(b)pyrrole	See indole			
1-Benzo(b)yrrole	See indole			
Benzothiazole		3256	See *Food Technol.*, May 1972	
1-Benzoxy-1-(2-methoxyethoxy) ethane	See benzyl methoxyethyl acetal			

Benzoylbenzene	See benzophenone			
Benzoyl eugenol	See eugenol benzoate			
Benzoylmethide	See acetophenone			
Benzoyl peroxide	Benzoyl superoxide		FDA §19 FDA §15	In specified cheeses, up to 0.0002% of milk Bleaching agent in flour; 1 part by weight of benzoyl peroxide mixed with not more than 6 parts by weight of one or more of potassium alum, calcium sulfate, magnesium carbonate, sodium aluminum sulfate, dicalcium phosphate, tricalcium phosphate, and sodium and calcium carbonate
Benzoyl phenyl carbinol	See benzoin			
Benzyl acetaldehyde	See 3-phenylpropionaldehyde			
Benzyl acetate	Acetic acid benzyl ester	2135	FDA §172.515	Synthetic flavor
Benzylacetic acid	See 3-phenylpropionic acid			
Benzyl acetoacetate	Benzyl acetyl acetate	2136	FDA §172.515	Synthetic flavor
Benzyl acetone	See 4-phenyl-3-buten-2-one			
Benzyl acetylacetate	See benzyl acetoacetate			
Benzyl alcohol	Phenylcarbinol	2137	FDA §172.515	Synthetic flavor
Benzyl alcohol benzoic ester	See benzyl benzoate			
Benzyl alcohol cinnamic ester	See benzyl cinnamate			
Benzyl alcohol, *p*-methoxy	See anisyl alcohol			
Benzyl alcohol, *p*-methoxy, acetate	See anisyl acetate			
Benzyl alcohol, *p*-methoxy, butyrate	See anisyl butyrate			
Benzyl alcohol, *p*-methoxy, phenylacetate	See anisyl phenylacetate			
Benzyl alcohol, *p*-methoxy, propionate	See anisyl propionate			
Benzyl benzenecarboxylate	See benzyl benzoate			
Benzyl benzoate	Benzyl benzene carboxylate	2138	FDA §172.515	Synthetic flavor
Benzyl butanoate	See benzyl butyrate			
Benzyl butyl alcohol	See α-propylphenethyl alcohol			
Benzyl butyl ether		2139	FDA §172.515	Synthetic flavor

REGULATORY STATUS OF DIRECT FOOD ADDITIVES (continued)

Material	Alternate Name [Source]	FEMA No.	Regulations	Limitations
Benzyl butyrate	Benzyl butanoate	2140	FDA §172.515	Synthetic flavor
Benzyl carbinol	See phenethyl alcohol			
Benzyl carbinyl acetate	See phenethyl acetate			
Benzyl carbinyl benzoate	See phenethyl benzoate			
Benzyl carbinyl butyrate	See phenethyl butyrate			
Benzyl carbinyl caproate	See phenethyl hexanoate			
Benzyl carbinyl cinnamate	See phenethyl cinnamate			
Benzyl carbinyl formate	See phenethyl formate			
Benzyl carbinyl hexanoate	See phenethyl hexanote			
Benzyl carbinyl 2-hydroxybenzoate	See phenethyl salicylate			
Benzyl carbinyl isobutyrate	See phenethyl isobutyrate			
Benzyl carbinyl isopentanoate	See phenethyl isovalerate			
Benzyl carbinyl methanoate	See phenethyl formate			
Benzyl carbinyl 3-methylbutanoate	See phenethyl isovalerate			
Benzyl carbinyl 2-methylbutyrate	See phenethyl 2-methylbutyrate			
Benzyl carbinyl 2-methylpropanoate	See phenethyl isobutyrate			
Benzyl carbinyl octanoate	See phenethyl octanoate			
Benzyl carbinyl phenylacetate	See phenethyl phenylacetate			
Benzyl carbinyl 3-phenylpropenoate	See phenethyl cinnamate			
Benzyl carbinyl propionate	See phenethyl propionate			
Benzyl carbinyl salicylate	See phenethyl salicylate			
Benzyl carbinyl tiglate	See phenethyl tiglate			
Benzyl carbinyl α-toluate	See phenethyl phenylacetate			
Benzyl carboxaldehyde	See phenylacetaldehyde			
Benzyl carboxylic acid	See phenylacetic acid			

Benzyl cinnamate	Benzyl β-phenylacrylate	2142	FDA §172.515	Synthetic flavor
Benzyl 2,3-dimethyl-2-butenoate	See benzyl 2,3-dimethylcrotonate			
Benzyl dimethyl carbinol	See α,α-dimethylphenethyl alcohol			
Benzyl dimethylcarbinyl acetate	See α,α-dimethylphenethyl acetate			
Benzyl dimethylcarbinyl butyrate	See α,α-dimethylphenethyl butyrate			
Benzyl dimethylcarbinyl formate	See α,α-dimethylphenethyl formate			
Benzyl 2,3-dimethylcrotonate	Benzyl methyl tiglate	2143	FDA §172.515	Synthetic flavor
2-Benzyl-4,5-dimethyl-1,3-dioxolane	See phenylacetaldehyde 2,3-butylene glycol acetal			
Benzyl dipropyl ketone	See 3-benzyl-4-heptanone			
Benzyl disulfide	Dibenzyl disulfide		FDA §172.515	Synthetic flavor
Benzyl dithiomethane	See methyl benzyl disulfide			
Benzyl ethanoate	See benzyl acetate			
Benzyl ether	See dibenzyl ether			
Benzyl ethyl alcohol	See 3-phenyl-1-propanol			
Benzyl ethyl ether		2144	FDA §172.515	Synthetic flavor
Benzyl formate	Formic acid benzyl ester	2145	FDA §172.515	Synthetic flavor
3-Benzyl-4-heptanone	Benzyl dipropyl ketone	2146	FDA §172.515	Synthetic flavor
Benzyl hydrosulfide	See benzyl mercaptan			
Benzyl *o*-hydroxybenzoate	See benzyl salicylate			
Benzylidencacetic acid	See cinnamic acid			
3-Benzylidene-2-butanone	See 3-methyl-4-phenyl-3-butene-2-one			
2-Benzylidene heptanol	See α-amylcinnamyl alcohol			
2-Benzylidenehexanal	See α-butylcinnamaldehyde			
2-Benzylideneoctanal	See α-hexylcinnamaldehyde			
Benzyl isoamyl alcohol	See α-isobutylphenethyl alcohol			
Benzyl isobutyl carbinol	See α-isobutylphenethyl alcohol			
Benzyl isobutyl ketone	See 4-methyl-1-phenyl-2-pentanone			
Benzyl isobutyrate		2141	FDA §172.515	Synthetic flavor
Benzyl isoeugenol	See isoeugenyl benzyl ether			
Benzyl isopeutanoate	See benzyl isovalerate			
Benzyl isovalerate		2152	FDA §172.515	Synthetic flavor
Benzyl isovalerianate	See benzyl isovalerate			

REGULATORY STATUS OF DIRECT FOOD ADDITIVES (continued)

Material	Alternate Name [Source]	FEMA No.	Regulation	Limitations
Benzyl β-ketobutyrate	See benzyl acetoacetate			
Benzyl ketone	See 1,3-diphenyl-2-propanone			
Benzyl mercaptan	α-Toluenethiol	2147	FDA §172.515	Synthetic flavor
Benzyl methanoate	See benzyl formate			
Benzyl methanol	See phenethyl alcohol			
Benzyl methoxyethyl acetal	Acetaldehyde benzyl β-methoxyethyl acetal	2148	FDA §172.515	Synthetic flavor
Benzyl 2-methoxy-4-propenylphenyl ether	See isoeugenyl benzyl ether			
Benzyl 3-methylbutanoate	See benzyl isovalerate			
Benzyl *trans* 2-methyl-2-butenoate		3330		
Benzyl methyl disulfide	See methyl benzyl disulfide			
Benzyl 2-methylpropanoate	See benzyl isobutyrate			
Benzyl methyl tiglate	See benzyl 2,3-dimethyl-crotonate			
Benzyl oxide	See dibenzyl ether			
Benzyl 3-oxobutanoate	See benzyl acetoacetate			
1-Benzyloxy-1-(β-methoxy) ethoxyethane	See benzyl methoxyethyl acetal			
1-Benzyloxy-2-methoxy-4-propenylbenzene	See isoeugenyl benzyl ether			
Benzyl phenylacetate		2149	FDA §172.515	Synthetic flavor
Benzyl β-phenylacrylate	See benzyl cinnamate			
Benzyl phenylformate	See benzyl benzoate			
Benzyl 3-phenylpropenoate	See benzyl cinnamate			
Benzyl propanoate	See benzyl propionate			
2-Benzyl-2-propanol	See α,α-dimethylphenethyl alcohol			
Benzyl propionate	Benzyl propanoate	2150	FDA §172.515	Synthetic flavor
2-Benzyl-2-propyl acetate	See α,α-dimethylphenethyl acetate			
2-Benzyl-2-propyl butyrate	See α,α-dimethylphenethyl butyrate			
Benzylpropyl carbinol	See α-propylphenethyl alcohol			
2-Benzyl-2-propyl formate	See α,α-dimethylphenethyl formate			

Benzyl salicylate	Benzyl *o*-hydroxybenzoate	2151	FDA §172.515	Synthetic flavor
Benzylthiol	See benzyl mercaptan			
Benzyl tiglate	See benzyl *trans*-2-methyl-2-butenoate			
Benzyl α-toluate	See benzyl phenylacetate			
Bergamol	See linalyl acetate			
Bergamot, oil	Bergamot orange, oil [*Citrus aurantium* (L.) subspecies *bergamia* Wright and Arn.]	2153	FDA §182.1	GRAS, natural flavor extractive
Bergamot orange, oil	See bergamot, oil [*C. aurantium* (L.) subspecies *bergamica* Wright and Arn.]			
BHA	Butylated hydroxyanisole; mixture of 2-*tert*-butyl-4-methoxyphenol and 3-*tert*-butyl-4-methoxyphenol	2183	FDA §182.1	GRAS, chemical preservative; total antioxidants, not over 0.02% of fat or oil content, including essential (volatile) oil content of food; product specifications apply
			FDA §121.1035	Antioxidant, alone or with BHT at max concentration of: 1. 1000 ppm BHA in active dry yeast 2. 2 ppm BHA in beverages and desserts prepared from dry mixes 3. 50 ppm BHA + BHT in dry breakfast cereals 4. 32 ppm BHA in dry, diced, glacéed fruit 5. 90 ppm BHA in dry mixes for beverages and desserts 6. 200 ppm BHA + BHT in emulsion stabilizers for shortenings 7. 50 ppm BHA + BHT in potato flakes 8. 10 ppm BHA + BHT in potato granules 9. 50 ppm BHA + BHT in sweet-potato flakes Product specifications apply
			FDA §121.1059	Antioxidant in chewing-gum base alone or with BHT and/or propyl gallate, up

REGULATORY STATUS OF DIRECT FOOD ADDITIVES (continued)

Material	Alternate Name [Source]	FEMA No.	Regulation	Limitations
				to 0.1% total antioxidant content
			FDA §172.515	Antioxidant in synthetic flavoring substances, up to 0.5% of essential oil content of flavoring substance
			MID	In unsmoked dry sausage, 0.003% BHA with 0.001% citric acid
			MID	To retard rancidity in unsmoked, dry sausage, up to 0.003%; in rendered fat, or combination of such fat and vegetable fat, up to 0.01%; in fat, BHA may be used with BHT, glycine, and propyl gallate, up to 0.02% of the combination
			MID	Antioxidant to retard rancidity in frozen, fresh pork sausage and freeze-dried meats, up to 0.01% based on fat content; also in combination with BHT and propyl gallate, to 0.02% in frozen, fresh pork sausage and up to 0.01% in freeze-dried meats
BHT	Butylated hydroxytoluene; 2,6-di-*tert*-butyl-4-methylphenol	2184	FDA §15.525	In enriched rice, up to 0.0033% by weight of enriched, parboiled rice
			FDA §182.1	GRAS, chemical preservative; total antioxidants, not over 0.02% of fat or oil content, including essential (volatile) oil content of food
			FDA §121.1034	Antioxidant, alone or in combination with BHA at max concentrations of: 1. 10 ppm BHT + BHA in potato granules 2. 50 ppm BHT + BHA in dry breakfast cereals

				3. 50 ppm BHT + BHA in potato and sweet-potato flakes 4. 200 ppm BHT + BHA in emulsion stabilizers for shortenings Product specifications apply
			FDA §121.1059	Antioxidant in chewing-gum base, alone or with BHA and/or propyl gallate, up to 0.1% total antioxidant content
			MID	Antioxidant in rendered animal fat or in combination with vegetable fat, up to 0.01% or up to 0.02% of a combination of BHT, BHA, and propyl gallate
			MID	Antioxidant to retard rancidity in frozen, fresh pork sausage and freeze-dried meats, up to 0.01% based on fat content; can be used in combination with BHA and propyl gallate, up to 0.02% in frozen, fresh pork sausage and up to 0.01% in freeze-dried meats
Biacetyl	See diacetyl			
Bible leaf	See costmary [*Chrysanthemum balsamita* (L.)]			
Biotin	*cis*-Tetrahydro-2-oxothienol[3,4-*d*] imidazoline-4-valeric acid		FDA §182.1	GRAS, nutrient and/or dietary supplement
Biphenyl		3129		Flavoring use at average max level of 2.0 ppm in beverages, frozen desserts, candies, baked goods, gelatins, and puddings
Biphenyl disulfide	See phenyl disulfide			
Birch, black, oil	See birch, sweet, oil [*Betula lenta* (L.)]			
Birch, sweet, oil	[*B. lenta* (L.)]	2154		
Birch, tar oil	[*B. pendula* Roth and related *Betula* spp.]	—	FDA §172.515	Synthetic flavor
Bisabolene		3331		
γ-Bisabolene	See bisabolene			
Bisabol myrrh	See opopanax			
Bis-(2,5-dimethyl-3-furyl) disulfide		3476		

REGULATORY STATUS OF DIRECT FOOD ADDITIVES (continued)

Material	Alternate Name [Source]	FEMA No.	Regulation	Limitations
Bis(2-furfuryl) disulfide	Furfuryl disulfide	3257	See *Food Technol.*, May 1972	
Bis(2-furfuryl) sulfide	Furfuryl sulfide	3258		See *Food Technol.*, May 1972
Bis(2-methyl-3-furyl) disulfide		3259		FEMA average max use levels — 0.1 ppm in baked goods, meat, meat sauces, soups, condiments, and pickles
Bis(2-methyl-3-furyl) tetrasulfide		3260		See *Food Technol.*, May 1972, for FEMA amul[a]
Bitter almond	See almond, bitter, oil			
Bitter almond oil, synthetic	See benzaldehyde			
Bitter ash, extract	See quassia, extract [*Picrasma excelsa* (Sw.) Planch. and *Quassia amara* (L.)]			
Bitter herb	See centaury [*Centaurium umbellatum* Gilib.]			
Bitter wood, extract	See quassia, extract [*Picrasma excelsa* (Sw.) Planch. and *Quassia amara* (L.)]			
Blackberry, bark, extract	[*Rubus* spp. of section Eubatus]	2155	FDA §172.510	Natural flavor
Black cutch, extract	See catechu, extract [*Acacia catechu* Willd.]			
Blackthorn berries	See sloe berries]*Prunus spinosa* (L.)]			
Blackthorn berries, extract	See sloe berries, extract]*P. spinosa* (L.)]			
Blackthorn berries, extract solid	See sloe berries, extract solid [*P. spinosa* (L.)]			
Black wattle	See mimosa			
Blätteralkohol	See 3-hexen-1-ol			
Bloodwort	See centaury [*Centaurium umbellatum* Gilib.]			
Blue, ultramarine	See ultramarine blue			
Bois de rose, oil	[*Aniba rosaeodora* Duckel]	2156	FDA §182.1	GRAS, natural flavor extractive

Boldo	See boldus leaves [*Peumus boldus* Mol.]			
Boldus, leaves	Boldo [*P. boldus* Mol.]	—	FDA §172.510	Natural flavoring in alcoholic beverages only
Boletic acid	See fumaric acid			
2-Bornanol	See borneol			
endo-2-Bornanol	See borneol			
exo-2-Bornanol	See isoborneol			
2-Bornanone	See *d*-camphor			
endo-2-Bornanyl	See bornyl formate			
Borneocamphor	See borneol			
Borneol	*d*-Camphanol; baros	2157	FDA §172.515	Synthetic flavor
Borneol (iso)	See isoborneol			
Borneol acetate	See bornyl acetate			
Borneol formate	See bornyl formate			
Bornyl acetate	Borneol acetate; terpeneless Siberian pine needle oil	2159	FDA §172.515	Synthetic flavor
Bornyl (iso) acetate	See isobornyl acetate			
D-Bornyl acetate	See bornyl acetate			
1-Bornyl acetate	See bornyl acetate			
exo-2-Bornyl acetate	See isobornyl acetate			
Bornyl alcohol	See borneol			
Bornyl ethanoate	See bornyl acetate			
Bornyl (iso) ethanoate	See isobornyl acetate			
Bornyl formate		2161	FDA §172.515	Synthetic flavor
Bornyl (iso) formate	See isobornyl formate			
D-Bornyl formate	See bornyl formate			
1-Bornyl formate	See bornyl formate			
exo-2-Bornyl formate	See isobornyl formate			
Bornyl isopentanoate	See bornyl isovalerate			
Bornyl (iso) isopentanoate	See isobornyl isovalerate			
Bornyl (iso) propionate	See isobornyl propionate			
Bornyl isovalerate	Bornyval	2165	FDA §172 .515	Synthetic flavor
Bornyl (iso) isovalerate	See isobornyl isovalerate			
Bornyl isovalerianate	See bornyl isovalerate			
Bornyl (iso) isovalerianate	See isobornyl isovalerate			
Bornyl methanoate	See bornyl formate			

REGULATORY STATUS OF DIRECT FOOD ADDITIVES (continued)

Material	Alternatate Name [Source]	FEMA No.	Regulation	Limitations
Bornyl (iso) 3-methylbutanoate	See isobornyl isovalerate			
Bornyl 3-methyl pentanoate	See bornyl isovalerate			
Bornyl *n*-pentanoate	See bornyl valerate			
exo-2-Bornyl propionate	See isobornyl propionate			
Bornyl valerate		2164	FDA §172.515	Synthetic flavor
Bornyl valerianate	See bornyl valerate			
Bornyval	See bornyl isovalerate			
Boronia, absolute	[*Boronia megastigma* Nees]	2167	FDA §172.510	
Boronia, flowers	[*B. megastigma* Nees]		FDA §172.510	Natural flavor
Bourbonal	See ethyl vanillin			
β-Bourbonene		—	FDA §172.515	Synthetic flavor
Brandy			FDA-GRAS	
Bromelin	Bromelain		MID	With water, salt, and monosodium glutamate to soften tissues of beef cuts; tissue weight gain, 3% max
Brominated vegetable oils		2168	FDA §121.4004	Stabilizer in citrus- and other fruit-flavored beverages at 15 ppm; *Food Chemicals Codex* specifications apply
Broom, extract	See genet, extract [*Spartium junceum* (L.)]			
Bryonia, root	[*Bryonia alba* (L.) and *B. dioica* Jacq.]	—	FDA § 172.510	Natural flavoring in alcoholic beverages only
Buchu leaves, oil	[*Barosma betulina* Bartl. and Wendl.; *B. crenulata* (L.) Hook.; *B. seratifolia* Willd.]	2169	FDA §172.510	Natural flavor
Buck bean leaves, extract	[*Menyanthes trifoliata* (L.)]	—	FDA §172.510	Natural flavor in alcoholic beverages only
Burtonite® V-7-E	See guar, gum [*Cyamopsis tetragonolobus* (L.)]			
Butadiene-styrene rubber			FDA §121.1059	Synthetic masticatory substance in chewing-gum base when additive is the basic polymer

Butanal	See butyraldehyde			
Butane	*n*-Butane		FDA §182.1	GRAS, miscellaneous and/or general-purpose food additive
1,4-Butanedicarboxylic acid	See adipic acid			
2,3-Butanedione	Diacetyl		FDA §182.1	GRAS, synthetic flavor
1,2-Butanedithiol	1,2-Dimercaptobutane	3528		See *Food Technol.*, February 1978
1,3-Butanedithiol	1,3-Dimercaptobutan	3529		See *Food Technol.*, February 1978
1-Butanethiol		3478		
Butanoic acid	See butyric acid			
Butanoic acid, *p*-methoxybenzyl ester	See anisyl butyrate			
Butanoic acid, 4-phenyl ethyl ester	See ethyl 4-phenylbutyrate			
1-Butanol	See *n*-butyl alcohol			
Butanol (iso)	See isobutyl alcohol			
2,3-Butanolone	See acetoin			
2-Butanone	Methyl ethyl ketone	2170	FDA §172.515	Synthetic fl avor
Butan-3-one-2-yl butanoate		3332		
2-Butanon-3-yl acetate	See acetyl methyl carbinyl acetate and 2-acetoxy-3-butanone			
trans-Butenedioic acid	See fumaric acid			
2-Butenoic acid, 2,3-dimethyl, benzyl ester	See benzyl 2,3-dimethylcrotonate			
trans-2-Butenoic acid, ethyl ester	See ethyl crotonate			
Butenyl α-cyclocitrylidenemethyl ketone	See allyl α-ionone			
Butter acids		2171	FDA §172.515	Synthetic flavor
Butter esters		2172	FDA §172.515	Synthetic flavor
Butter of zinc	See zinc chloride			
Butter starter distillate		2173	FDA §172.515	Synthetic flavor
Butyl acetate		2174	FDA §172.515	Synthetic flavor
Butyl (iso) acetate	See isobutyl acetate			
sec-Butyl acetic acid	See 3-methylpentanoic acid			
Butyl acetoacetate		2176	FDA §172.515	Synthetic flavor
Butyl (iso) acetoacetate	See isobutyl acetoacetate			
Butyl acetyl propionate	See butyl levulinate			
3-Butylacrolein	See *trans*-2-heptenal			

REGULATORY STATUS OF DIRECT FOOD ADDITIVES (continued)

Material	Alternate Name [Source]	FEMA No.	Regulation	Limitations
n-Butyl alcohol	1-Butanol	2178	FDA §172.515	Synthetic flavor
			FDA §8.390 and §8.6200	Diluent in inks for marking gum and confectionery, for marking food supplements in tablet form, and for branding pharmaceutical forms; product specifications apply
Butyl (iso) alcohol	See isobutyl alcohol			
Butyl aldehyde	See butyraldehyde			
Butyl (iso) aldehyde	See isobutyraldehyde			
Butylamine	*dl*-(3-Amino-3-carboxypropyl) dimethyl sulfonium; vitamin U	3130		
Butyl (iso) angelate	See isobutyl angelate			
Butyl anthranilate		2181	FDA §172.515	Synthetic flavor
Butyl (iso) anthranilate	See isobutyl anthranilate			
Butylated hydroxyanisole	See BHA			
Butylated hydroxytoluene	See BHT			
Butyl (iso) benzoate	See isobutyl benzoate			
Butyl benzyl ether	See benzyl butyl ether			
Butyl butanoate	See butyl butyrate			
Butyl (iso) butanoate	See isobutyl butyrate			
2-Butyl-2-butenal		3392		
Butyl butyrate		2186	FDA §172.515	Synthetic flavor
γ-*n*-Butyl-γ-butyrolactone	See γ-octalactone			
Butyl butyryllactate	Lactic acid, butyl ester, butyrate	2190	FDA §172.515	Synthetic flavor
Butyl caproate	See butyl hexanoate			
Butyl (iso) caproate	See isobutyl hexanoate			
n-Butyl carbinol	See amyl alcohol			
Butyl (iso) carbinol	See isoamyl alcohol			
Butyl (iso) carbylamine	See isopentylamine			
α-Butylcinnamaldehyde		2191	FDA §172.515	Synthetic flavor
Butyl cinnamate		2192	FDA §172.515	Synthetic flavor
Butyl (iso) cinnamate	See isobutyl cinnamate			

Butylcinnamic aldehyde	See α-butylcinnamaldehyde			
2-*sec*-Butylcyclohexanone		3261		See *Food Technol.*, May 1972
o-*sec*-Butylcyclohexanone	See 2-*sec*-butylcyclohexanone			
Butyl 2-decenoate	Butyl decylenate	2194	FDA §172.515	Synthetic flavor
Butyl decylenate	See butyl 2-decenoate			
2-Butyl-5- or 6-keto-1,4-dioxane		2204		
Butyl dodecanate	See butyl laurate			
Butyl dodecylate	See butyl laurate			
Butyl ethyl carbinol	See 3-heptanol			
1,3-Butylene glycol			FDA §121.1176	Solvent for natural and synthetic flavoring substances; product specifications apply
Butyl ethanoate	See butyl acetate			
Butyl (iso) ethanoate	See isobutyl acetate			
n-Butyl ethyl carbinol	See 3-heptanol			
sec-Butyl ethyl ether		3131		Flavoring in beverages, frozen desserts, candies, and baked goods at average max level of 1.0 ppm
Butyl ethyl ketone	See 3-heptanone			
Butyl ethyl malonate		2195	FDA §172.515	Synthetic flavor
Butyl formate		2196	FDA §172.515	Synthetic flavor
Butyl (iso) formate	See isobutyl formate			
Butyl (iso) 2-furanpropionate	See isobutyl 2-furanpropionate			
Butyl (iso) furylpropionate	See isobutyl 2-furanpropionate			
Butyl heptanoate		2199	FDA §172.515	Synthetic flavor
Butyl (iso) heptanoate	See isobutyl heptanoate			
Butyl heptoate	See butyl heptanoate			
Butyl (iso) heptoate	See isobutyl heptanoate			
Butyl heptylate	See butyl heptanoate			
Butyl hexanoate		2201	FDA §172.515	Synthetic flavor
Butyl (iso) hexanoate	See isobutyl hexanoate			
Butyl *p*-hydroxybenzoate	Butyl Parasept®	2203	FDA §172.515	Synthetic flavor
Butyl (iso) *o*-hydroxybenzoate	See isobutyl salicylate			
n-Butyl *o*-hydroxybenzoate	See butyl salicylate			

REGULATORY STATUS OF DIRECT FOOD ADDITIVES (continued)

Material	Alternate Name [Source]	FEMA No.	Regulations	Limitations
4-*n*-Butyl-4-hydroxybutyric acid lactone	See γ-octalactone			
4-Butyl-4-hydroxyoctanoic acid lactone	See 4,4-dibutyl-γ-butyrolactone			
5-*n*-Butyl-5-hydroxypentanoic acid lactone	See hydroxynonanoic acid, δ-lactone			
Butyl (2-hydroxyphenyl)formate	See butyl salicylate			
Butyl 2-hydroxypropanoate	See butyl lactate			
n-Butylideneacetone	See 3-hepten-2-one			
3-Butylidenephthalide		3333		
Butyl isobutyrate		2188	FDA §172.515	Synthetic flavor
Butyl (iso) isobutyrate	See isobutyl isobutyrate			
Butyl isopentanoate	See butyl isovalerate			
Butyl isovalerate		2218	FDA §172.515	Synthetic flavor
Butyl isovalerianate	See butyl isovalerate			
Butyl 3-ketobutanoate	See butyl acetoacetate			
Butyl (iso) 3-ketobutanoate	See isobutyl acetoacetate			
Butyl 3-ketobutyrate	See butyl acetoacetate			
Butyl (iso) 3-ketobutyrate	See isobutyl acetoacetate			
Butyl 4-ketovalerate	See butyl levulinate			
Butyl lactate		2205	FDA §172.515	Synthetic flavor
Butyl laurate	Butyl dodecanoate	2206	FDA §172.515	Synthetic flavor
Butyl levulinate		2207	FDA §172.515	Synthetic flavor
n-Butyl mercaptan	See 1-butanethiol			
Butyl methanoate	See butyl formate			
Butyl (iso) methanoate	See isobutyl formate			
2-*tert*-Butyl-4-methoxyphenol	See butylated hydroxyanisole			
3-*tert*-Butyl-4-methoxyphenol	See butylated hydroxyanisole			
2-Butyl (iso)-3-methoxypyrazine	See 2-isobutyl-3-methoxypyrazine			

Butyl 3-methylbutanoate	See butyl isovalerate			
Butyl (iso) *cis*-2-methyl-2-butenoate	See isobutyl angelate			
n-Butyl 2-methylbutyrate		3393		
2-Butyl (iso)-5-methylcyclohexanal	See *d*-neo-menthol			
Butyl (iso) methyl ketone	See 4-methyl-2-pentanone			
Butyl 2-methylpropanoate	See butyl isobutyrate			
Butyl (iso) 2-methylpropanoate	See isobutyl isobutyrate			
2-Butyl (iso)-3-methylpyrazine	See 2-isobutyl-3-methylpyrazine			
Butyl octadecanoate	See butyl stearate			
Butyl 3-oxobutanoate	See butyl acetoacetate			
Butyl (iso) 3-oxobutanoate	See isobutyl acetoacetate			
Butyl 4-oxopentanoate	See butyl levulinate			
Butyl paraben	See butyl *p*-hydroxybenzoate			
Butyl Parasept	See butyl *p*-hydroxybenzoate			
Butyl peutanoate	See butyl valerate			
α-Butyl (iso) phenethyl alcohol	See α-isobutylphenethyl alcohol			
Butyl phenylacetate		2209	FDA §172.515	Synthetic flavor
Butyl (iso) phenylacetate	See isobutyl phenylacetate			
α-Butyl-β-phenylacrolein	See α-butylcinnamaldehyde			
n-Butyl phenylacrylate	See butyl cinnamate			
Butyl β-phenylacrylate	See butyl cinnamate			
Butyl 3-phenylpropenoate	See butyl cinnamate			
3-*n*-Butylphthalide		3334		
Butyl propanoate	See butyl propionate			
Butyl (iso) propanoate	See isobutyl propionate			
Butyl propionate		2211	FDA §172.515	Synthetic flavor
Butyl (iso) propionate	See isobutyl propionate			
Butyl salicylate		—		
Butyl (iso) salicylate	See isobutyl salicylate			
Butyl sebacate	See dibutyl sebacate			
Butyl stearate	Butyl octadecanoate	2214	FDA §172.515	Synthetic flavor
Butyl sulfide	Dibutyl sulfide	2215	FDA §172.515	Synthetic flavor
2-*sec*-Butylthiazole	See 2-(1-methylpropyl)thiazole			

REGULATORY STATUS OF DIRECT FOOD ADDITIVES (continued)

Material	Alternate Name [Source]	FEMA No.	Regulation	Limitations
2-Butyl (iso) thiazole	See 2-isobutyl thiazole			
Butyl thiobutane	See butyl sulfide			
Butyl α-toluate	See butyl phenylacetate			
Butyl 10-undecenoate		2216	FDA §172.515	Synthetic flavor
Butyl undecylenate	See butyl 10-undecenoate			
Butyl valerate		2217	FDA §172.515	Synthetic flavor
Butyl valerianate	See butyl valerate			
5-*n*-Butyl-δ-valerolactone	See hydroxynonanoic acid, δ-lactone			
Butyraldehyde	Butanal	2219	FDA §172.515	Synthetic flavor
Butyraldehyde (iso)	See isobutyraldehyde			
n-Butyric acid	Butanoic acid	2221	FDA §182.1	GRAS, synthetic flavor
Butyric (iso) acid	See isobutyric acid			
Butyric acid, benzyl ester	See benzyl butyrate			
Butyric acid, *p*-methoxybenzyl ester	See anisyl butyrate			
Butyric acid, 3-phenyl-2-propen-1-yl ester	See cinnamyl butyrate			
Butyric aldehyde	See butyraldehyde			
Butyric (iso) aldehyde	See isobutyraldehyde			
Butyrin	See (tri-)butyrin			
(tri-)Butyrin	Butyrin; glycerol trbutyrate; glyceryl tributyrate; tributyrin	2223	FDA §172.515	Synthetic flavor
Butyroin	See 5-hydroxy-4-octanone		FDA §182.1	GRAS, synthetic flavor and adjuvant
γ-Butyrolactone	See 4-hydroxybutanoic acid lactone			
Butyrone	See 4-heptanone			
Butyryl acetyl	See 2,3-hexanedione			
Buxine®	See α-amylcinnamaldehyde			
C-8	See octanoic acid			
Cacao	Cocoa [*Theobroma cacao* (L.)]		FDA §182.1	GRAS, natural flavor extractive
Cachou, extract	See catechu, extract [*Acacia catechu* Willd.]			
Cactus root, extract	See yucca, Mohave, extract [*Yucca* spp.]			

Cade oil	Juniper tar; empyreumatic wood oil [*Juniperus oxycedrus* (L.)]		FDA §172.515	Synthetic flavor
Cadinene		—	FDA §172.515	Synthetic flavor
Caffeine	1,3,7-Trimethylxanthine	2224	FDA §182.1	GRAS, cola-type beverages, up to 0.02%
Cajeput, oil	[*Melaleuca leucadendron* (L.) other *Melaleuca* spp.]	2225	FDA §172.510	Natural flavors
Cajeputene	See *d*-limonene			
Cajeputol	See eucalyptol			
Calamus	[*Acorus calamus* (L.)]	—	FDA §172.510	
Calcium acetate		2228	FDA §182.1	GRAS, sequestrant
Calendula	See marigold, pot [*Calendula officinalis* (L.)]			
Calcium alginate	See alginates — sodium, calcium, and ammonium salts			
Calcium ascorbate	Ascorbic acid calcium salt		FDA §182.1	GRAS, chemical preservative
Calcium caprate	Calcium salt of capric acid		FDA §121.1071	Binder, emulsifier, and anticaking agent; complying with §121.1070; product specifications apply
Calcium caprylate	Calcium salt of caprylic acid		FDA §121.1071	Binder, emulsifier, and anticaking agent; complying with §121.1070; product specifications apply
Calcium carbonate	Carbamic acid calcium salt		FDA §182.1	GRAS, nutrient and/or dietary supplement
			FDA §182.1	GRAS, miscellaneous and/or general-purpose food additive
Calcium caseinate			FDA §20	In frozen desserts
Calcium chloride			FD §18.520	In evaporated milk, up to 0.1% by weight of finished product
			FDA §182.1	GRAS, sequestrant
			FDA §182.1	GRAS, miscellaneous and/or general-purpose food additive
Calcium citrate			FDA §182.1	GRAS, nutrient and/or dietary supplement
			FDA §182.1	GRAS, sequestrant
			FDA §182.1	GRAS, miscellaneous and/or general-purpose food additive
Calcium diacetate			FDA §182.1	GRAS, sequestrant

REGULATORY STATUS OF DIRECT FOOD ADDITIVES (continued)

Material	Alternate Name [Source]	FEMA No.	Regulation	Limitations
Calcium disodium EDTA	Calcium disodium ethylenediaminetetraacetate; calcium disodium (ethylenedinitrilo)-tetraacetate		FDA §121.1017	Alone as food additive: 1. 33 ppm max in canned, carbonated soft drinks 2. 110 ppm max in canned white potatoes 3. 340 ppm max in canned cooked clams 4. 275 ppm max in canned cooked crabmeat 5. 25 ppm max in distilled alcoholic beverages 6. 75 ppm max in nonstandardized dressings 7. 310 ppm max in canned, cooked dried lima beans 8. 25 ppm max in fermented malt beverages 9. 75 ppm max in French dressing 10. 75 ppm max in mayonnaise 11. 200 ppm max in canned, cooked mushrooms 12. 75 ppm max in oleomargarine 13. 100 ppm max in pecan pie filling 14. 220 ppm max in pickled cabbage 15. 220 ppm max in pickled cucumbers 16. 100 ppm max in potato salad 17. 800 ppm max in processed, dry pinto beans 18. 75 ppm max in salad dressings

			19. 100 ppm max in sandwich spreads 20. 75 ppm max in sauces 21. 250 ppm max in canned, cooked shrimp 22. 60 ppm max in spice extractives in soluble carriers 23. 100 ppm max in artificially flavored lemon and orange spreads In combination with disodium EDTA as food additive: 1. 75 ppm max in nonstandardized dressings 2. 75 ppm max in French dressing 3. 75 ppm max in mayonnaise 4. 75 ppm max in salad dressing 5. 100 ppm max in sandwich spread 6. 75 ppm max in sauces Product specifications apply
Calcium gluconate		FDA §182.1 FDA §182.1	GRAS, sequestrant GRAS, miscellaneous and/or general-purpose food additive
Calcium glycerophosphate	Glycerophosphoric acid calcium salt	FDA §182.1	GRAS, nutrient and/or dietary supplement
Calcium hexametaphosphate		FDA §182.1	GRAS, sequestrant
Calcium hydroxide	Lime	FDA §182.1	GRAS, miscellaneous and/or general-purpose food additive
		FDA §8.303	Caramelization of caramel for food use
Calcium iodate		FDA §17	In bakery products, alone or with potassium bromate,potassium iodate, and/or calcium peroxide, up to 0.0075% by weight of flour
Calcium lactate		FDA §182.1	GRAS, miscellaneous and/or general-purpose food additive
Calcium lactobionate	Calcium salt of lactobionic acid	FDA §121.1162	Firming agent in dry pudding mixes; product specifications apply
Calcium laurate	Calcium salt of lauric acid	FDA §121.1071	Binder, emulsifier, and anticaking agent; complying with §121.1070; product specifications apply

REGULATORY STATUS OF DIRECT FOOD ADDITIVES (continued)

Material	Alternate Name [Source]	FEMA No.	Regulation	Limitations
Calcium myristate	Calcium salt of myristic acid		FDA §121.1071	Binder, emulsifier, and anticaking agent; complying with §121.1070; product specifications apply
Calcium oleate	Calcium salt of oleic acid		FDA §121.1071	Binder, emulsifier, and anticaking agent; complying with §121.1070; product specifications apply
Calcium oxide			FDA §182.1	GRAS, nutrient and/or dietary supplement
			FDA §182.1	GRAS, miscellaneous and/or general-purpose food additive
Calcium palmitate	Calcium salt of palmitic acid		FDA §121.1071	Binder, emulsifier, and anticaking agent, complying with §121.1070; product specifications apply
Calcium pantothenate	Pantothenic acid calcium salt		FDA §182.1	GRAS, nutrient and/or dietary supplement
			FDA §121.1037	Calcium chloride double salt of calcium pantothenate; special dietary uses; product specifications apply
Calcium peroxide	Calcium dioxide		FDA §17	In bakery products, alone or with potassium bromate, potassium iodate, and/or calcium iodate, up to 0.0075% by weight of flour
Calcium phosphate, monobasic	Acid calcium phosphate; monocalcium phosphate		FDA §182.1	GRAS, sequestran
Calcium phosphate, mono-, di-, tribasic			FDA §182.1	GRAS, nutrient and/or dietary supplement; GRAS, miscellaneous and/or general-purpose food additive
Calcium phytate			FDA §182.1	GRAS, sequestrant
Calcium propionate	Propionic acid calcium salt		FDA §17	In bakery products, alone or with sodium propionate, up to 0.32% by weight of flour used for breads (enriched, milk, and raisin breads) and up

			to 0.38% of flour used in whole-wheat bread
		FDA §182.1	GRAS, chemical preservative
		MID	To retard mold growth in pizza crust, up to 0.32% alone or in combination with sodium propionate, based on weight of flour used; product specifi-flapply
Calcium pyrophosphate	Calcium diphosphate	FDA §182.1	GRAS, nutrient and/or dietary supplement
Calcium saccharin		FDA §182.1	GRAS, nonnutritive sweetener
Calcium salts		FDA §17.2	In enriched bread; each pound of finished food to contain 300 to 880 mg of calcium
Calcium salts of fatty acids		FDA §121.1071	Binder, emulsifier, and anticaking agent; complying with §121.1070; product specifications apply
Calcium silicate		FDA§182.1	GRAS, anticaking agent in baking powder, up to 5%; GRAS, anticaking agent in table salt, up to 2%
		FDA §121.1135	Anticaking agent in food; product specifications apply
Calcium sorbate		FDA §182.1	GRAS, chemical preservative
Calcium stearate	Calcium salt of stearic acid	FDA §22.8 and §29.9	In vanilla powder and vanilla-vanillin powder
		FDA §121.1071	Binder, emulsifier, and anticaking agent; complying with §121.1070; product specifications apply
		FDA §121.1099	In defoaming agents in processing beet sugar and yeast, conforming to §121.1071
Calcium stearyl-2-lactylate		FDA §121.1047	1. Dough conditioner in yeast-leavened bakery products and prepared mixes for yeast-leavened bakery products, 0.5% max by weight of flour 2. Whipping agent in liquid and frozen egg whites, 0.05% max; whipping agent in dried egg white, 0.5% max;

REGULATORY STATUS OF DIRECT FOOD ADDITIVES (continued)

Material	Alternatate Name [Source]	FEMA No.	Regulation	Limitations
				whipped vegetable oil topping, 0.3% max by weight of finished vegetable oil topping
				Product specifications apply
Calcium sulfate			FDA §51.990	In specified canned vegetables
			FDA §17	In bakery products, alone or with dicalcium phosphate and/or ammonium chloride, 0.25% max by weight of flour
			FDA §20	In frozen desserts
			FDA §15	Bleaching agent in flour, not more than 6% by weight, alone or in combination with potassium alum, magnesium carbonate, sodium aluminum sulfate, dicalcium phosphate, tricalcium phosphate, starch, and/or calcium carbonate, and with one part by weight of benzoyl peroxide
			FDA §19	In specified cheeses, alone or in comination, up to 6 times the weight of benzoyl peroxide used
			FDA §182.1	GRAS, nutrient and/or dietary supplement
Calendula	See Marigold, pot [*Calendula officinalis*(L.)]			
Calgon®	See sodium hexametaphosphate			
Calumba root	[*Jateorhiza palmata*(Lam) Miers]	—	FDA §172.510	Natural flavoring in alcoholic beverages only
Calumba root, extract	[*J. palmata*(Lam.) Miers]	—	FDA §172.510	
Camomile, flower, English, oil	Chamomile [*Anthemis nobilis*(L.)]	2272	FDA §182.1	GRAS, natural flavor extractive
Camomile, flower, Hungarian, oil	Chamomile [*Matricaria chamomilla* (L.)]	2273	FDA §182.1	GRAS, natural flavor extractive

Camomile, flower, Roman, extract	Chamomile [*A. nobilis*(L.)]	2274	FDA §182.1	GRAS, natural flavor
Camomile, flower, Roman, oil	Chamomile [*A. nobilis*(L.)]	2275	FDA §182.1	GRAS, natural flavor extractive
d-Camphanol	See borneol			
endo-2-Camphanol	See borneol			
exo-Camphanol	See isoborneol			
2-Camphanol	See borneol			
2-Camphanone	See *d*-camphor			
2-Camphanyl acetate	See bornyl acetate			
exo-2-Camphanyl acetate	See isobornyl acetate			
2-Camphanyl formate	See bornyl formate			
exo-2-Camphanyl formate	See isobornyl formate			
exo-2-Camphanyl propionate	See isobornyl propionate			
endo-2-Camphanyl valerate	See bornyl valerate			
Camphene	2,2-dimethyl-3-methylenenorborane	2229	FDA §172.515	Synthetic flavor
Camphol	See borneol			
iso-Camphol	See isoborneol			
Campholenic aldehyde	See 2,2,3-trimethylcyclopent-3-en-1-ylacetaldehyde			
d-Camphor		2230	FDA §172.515	Synthetic flavor
Camphor, Japanese, white, oil	[*Cinnamomum camphora*(L.) Nees and Eberm.]	2231	FDA §172.510	Natural flavor; safrole-free
Cananga, oil	[*Cananga odorata* Hook f. and Thoms.]	2232	FDA §182.1	GRAS, natural flavor extractive
Candelilla wax	[Wax from stems and branches of *Euphorbia cerifera*]	3479		
Canthaxanthin	*β*-Carotene-4,4′-dione		FDA §8.326	Food color; not to exceed 30 mg per lb of solid or semisolid food or per pt of liquid food; may not be used to color standardized foods unless authorized by the food standard; product specifications apply
Capers	[*Capparis spinosa*(L.)]		FDA §182.1	GRAS, natural flavor
Capraldehyde	See decanal			
Capraldehyde dimethyl acetal	See decanal dimethyl acetal			

REGULATORY STATUS OF DIRECT FOOD ADDITIVES (continued)

Material	Alternate Name [Source]	FEMA No.	Regulation	Limitations
Capric acid	Decanoic acid	2364	FDA §121.1070	Lubricant, binder, and defoaming agent in foods; product specifications apply
Capric aldehyde	See decanal			
Caprinaldehyde	See decanal			
Caproaldehyde	See hexanal			
Caproic acid	See hexanoic acid			
Caproic alcohol	See hexyl alcohol			
Caproic aldehyde	See hexanal			
γ-Caprolactone	See γ-hexalactone			
δ-Caprolactone	See δ-hexalactone			
Capronic ether, absolute	See ethyl hexanoate			
Caproylethanol	See 3-octanon-1-ol			
Capryl alcohol	See 1-octanol			
sec-Capryl alcohol	See 2-octanol			
Capryl alcohol, secondary	See 2-octanol			
Caprylaldehyde	See *n*-octanal			
Caprylaldehyde dimethyl acetal	See octanal dimethyl acetal			
Caprylic acid	See octanoic acid			
Caprylic alcohol	See 1-octanol			
sec-Caprylic alcohol	See 2-octanol			
Caprylic aldehyde	See octanal			
Capsaicin	See *N*-(4-hydroxy-3-methoxybenzyl)-8-methyl-6-nonenamide			
Capsicum, extract	Cayenne pepper, red pepper, and paprika [*Capsicum frutescens* (L.) or *C. annuum* (L.)]	2233	FDA §182.1	GRAS, natural flavor and natural flavor extractive
Capsicum, oleoresin	Cayenne pepper, red pepper, and paprika [*Capsicum frutescens* (L.) or *C. annuum* (L.)]	2234	FDA §182.1	GRAS, natural flavor and natural flavor extractive
Caramel			FDA §182.1	GRAS, miscellaneous and/or general-purpose food additive
			FDA §8.303	Food color; product specifications apply

Caramel color		2235	FDA §121.303	
Caraway	[*Carum carvi*(L.)]	2236	FDA §182.1	GRAS, natural flavor and natural flavor extractive
Caraway, black	Black cumin [*Nigella sativa*(L.)]	2237	FDA §182.1	GRAS, natural flavor
Caraway, oil	[*Carum carvi*(L.)]	2238	FDA §182.1	Natural flavor
Carbohydrase	[*Aspergillus niger*]		FDA §121.1233	Removal of visceral mass in clam processing and in removal of shell from edible tissue in shrimp processing; product specifications apply
	[*Rhizopus oryzae*]		FDA §121.1165	In production of dextrose from starch; product specifications apply
Carbolic acid	See phenol			
Carbonates			FDA §8.305	Adjusting pH of food-grade annatto extract
Carbon dioxide			FDA §182.1	GRAS, miscellaneous and/or general-purpose food additive
			FDA §121.1065	With octafluorocyclobutane as propellant and aerating agent for foamed or sprayed food products
			FDA §121.1180	With chloropentafluoroethane as propellant and aerating agent for foamed or sprayed food products
Carbon dioxide, solid	Dry ice		MID	For cooling during chopping and packaging of meat
Carboxybenzene	See benzoic acid			
2-Carboxyglutaconic acid	See aconitic acid			
Carboxymethylcellulose	See sodium carboxymethylcellulose			
Cardamon	[*Elletaria cardamomum* (L.) Maton]	2240	FDA §182.1	Natural flavor
Cardamom seed	[*E. cardamomum* (L.) Maton]	2241	FDA §182.1	GRAS, natural flavor and natural flavor extractive
Carmine	[*Coccus cacti*(L.)]	2242	FDA §8.317	Food color; product specifications apply
Carminic acid (see also cochineal extract)	Cochineal extract		FDA §8.317 and §8.6009	In production of aluminum or calcium/aluminum lakes of F&D dyes
Carnauba wax	[*Copernicia prunifera* (Muell.) H. E. Moore]		FDA §182.1	GRAS, miscellaneous and/or general-purpose food additive
Carob bean, extract	St. John's bread [*Ceratonia siliqua* (L.)]	2243	FDA §182.1	GRAS, natural flavor extractive

REGULATORY STATUS OF DIRECT FOOD ADDITIVES (continued)

Material	Alternate Name [Source]	FEMA No.	Regulations	Limitations
Carob bean gum	Locust bean, gum; locust gum [*Ceratonia siliqua* (L.)]	2658	FDA §182.1	GRAS, stabilizer
Carotene			FDA §182.1	GRAS, nutrient and/or dietary supplement
β-Carotene			FDA §8.304	Food color; product specifications apply
Carrageenan	Chondurs extract [*Chondrus crispus* (L.) Stackhouse; *C. ocellatus*; *Eucheuma spinosum*; *E. cottonii*; *Gigartina stellata*; and other species]		FDA §121.1066	Emulsifier, stabilizer, or thickener in foods; product specifications apply
			FDA §182.1	GRAS, stabilizer
			MID	To extend and stabilize breading mix and sauces
Carrageenan, salts of			FDA §121.1067	Emulsifier, stabilizer, or thickener in foods; product specifications apply
Carrageenan with polysorbate 80			FDA §121.1193	Polysorbate 80 at 5% to facilitate separation of sheeting carrageenan and its salts from drying rolls; carrageenan and its salts so produced — emulsifier, stabilizer or thickener in foods; 500 ppm max of polysorbate 80 in final food; product specifications apply
Carrageen, extract	See Irish moss, extract			
Carragheen	See carrageenan			
Carrot oil	[*Daucus carota* (L.)]	2244	FDA §182.1	GRAS, natural flavor extractive
			FDA §8.324	Food color; product specifications apply
Carvacrol	2-*p*-Cymenol	2245	FDA §172.515	Synthetic flavor
Carvacryl ethyl ether	2-Ethoxy-*p*-cymene	2246	FDA §172.515	Synthetic flavor
Carvene	See *d*-limonene			
Carveol	*p*-Mentha-6,8-dien-2-ol	2247	FDA §172.515	Synthetic flavor
iso-Carveol	See *p*-mentha-1,8-dien-7-ol			
Carvol	See carvone			

Carvomenthenol	See *p*-menthan-2-ol	—		
4-Carvomenthenol	1-*p*-*Menthen-4-ol*; 4-terpinenol	2248	FDA §172.515	Synthetic flavor
Carvomenthone	See *p*-menthan-2-one			
Carvone, *d*- or *l*-	Carvol	2249	FDA §182.1	GRAS, synthetic flavor
cis-Carvone oxide		—	FDA §172.515	Synthetic flavor
Carvyl acetate		2250	FDA §172.515	Synthetic flavor
Carvyl propionate		2251	FDA §172.515	Synthetic flavor
Caryolan-1-ol	See caryophyllene alcohol			
β-Caryophyllene		2252	FDA §172.515	Synthetic flavor/adjuvant
Caryophyllene acetate	See caryophyllene alcohol acetate			
Caryophyllene alcohol		—	FDA §172.515	Synthetic flavor/adjuvant
Caryophyllene alcohol acetate		—	FDA §172.515	Synthetic flavor/adjuvant
Cascara sagrada, bitterless, extract	[*Rhamnus purshiana* D.C.]	2253	FDA §172.510	Natural flavor
Cascarilla bark, extract	[*Croton cascarilla* Benn.; *C. eluteria* Benn.]	2255	FDA §182.1	Natural flavor
Casein			FDA §20	In frozen desserts
Cashoo, extract	See catechu, extract [*Acacia catechu* Willd.]			
Cassia bark, Chinese	[*Cinnamomum cassia* Blume]	2257	FDA §182.1	GRAS, natural flavor extractive
Cassia bark, extract	[*C. cassia* Blume]	2257	FDA §182.1	Natural flavor
Cassia bark, oil	[*C. cassia* Blume]	2258	FDA §182.1	Natural flavor
Cassia bark, Padang or Batavia	[*C. burmanni* Blume]		FDA §182.1	GRAS, natural flavor extractive
Cassia bark, Saigon	[*C. loureirii* Nees]		FDA §182.1	GRAS, natural flavor extractive
Cassia buds	[*C. cassia* Blume]	2259		
Cassia, Chinese	[*C. cassia* Blume]	2256	FDA §182.1	GRAS, natural flavor
Cassia flowers	See cassia buds [*C. cassia* Blume]			
Cassia, Padang or Batavia	[*C. burmanni* Blume]		FDA §182.1	GRAS, natural flavor
Cassia, Saigon	[*C. loureirii* Nees]		FDA §182.1	GRAS, natural flavor
Cassie, absolute	[*Acacia farnesiana* (L.) Willd.]	2260	FDA §172.510	
Cassie flowers	[*A. farnesiana* (L.) Willd.]		FDA §172.510	Natural flavor
Cassi ketone	See anisyl acetate			
Cassis	See currant buds, black, absolute [*Ribes nigrum* (L.)]			
Castoreum, extract	[*Castor fiber* (L.) and *C. canadensis* Kuhl]	2261	FDA §182.1	GRAS, natural flavor extractive

REGULATORY STATUS OF DIRECT FOOD ADDITIVES (continued)

Material	Alternate Name [Source]	FEMA No.	Regulation	Limitations
Castoreum, liquid	[*C. fiber* (L.) and *C. canadensis* Kuhl]	2262		
Castor oil	[*Ricinus communis* (L.)]	2263	FDA §121.1028	Release agent and antisticking agent in hard candy production, 500 ppm max; component of protective coatings in vitamin and mineral tablets; product specifications apply
			FDA §172.510	Natural substance used in conjunction with flavors
			FDA §8.390 and §8.6200	Diluent in inks for marking gum, confectionery, and food supplements in tablet form; inks for marking fruits and vegetables; mixtures for coloring eggshells and ingested and externally applied drugs; inks for branding pharmaceutical forms; in foods, 500 ppm max; product specifications apply
Catalase	See bacterial catalase			
Catechu, black, extract	[*Acacia catechu* Willd.]	2264	FDA §172.510	Natural flavor
Catechu, black, powder	[*A. catechu* Willd.]	2265	FDA §172.510	
Catechu, pale	See gambir			
Caustic soda	See sodium hydroxide			
Cayenne pepper	[*Capsicum annuum* (L.) var. *longum* Sendt]	2266	FDA §182.1	Natural flavor
C-8 Dimethylacetal	See octanal dimethyl acetal			
Cedar leaf, oil	[*Thuja occidentalis* (L.)]	2267	FDA §172.510	
Cedar, white, leaves and twigs	Arborvitae [*T. occidentalis* (L.)]		FDA §172.510	Natural flavor; thujone-free
Cedarwood oil, alcohols		—	FDA §172.515	Synthetic flavor
Cedarwood oil, terpenes		—	FDA §172.515	Synthetic flavor
Cedro, oil	See lemon, oil, terpeneless [*Citrus limon* (L.) Burm. f.]			
Celery ketone	See 3-methyl-5-propyl-2-cyclohexen-1-one			

Celery seed	[*Apium graveolens*(L.)]	2268	FDA §182.1	GRAS, natural flavor and natural flavor extractive
Celery seed, extract	[*A. graveolens*(L.)]	2269	FDA §182.1	Natural flavor
Celery seed, extract solid	[*A. graveolens*(L.)]	2270	FDA §182.1	Natural flavor
Celery seed, oil	[*A. graveolens*(L.)]	2271	FDA §182.1	Natural flavor
Cellulase	[*Aspergillus niger*]		FDA §121.1233	To remove visceral mass in clam processing; product specifications apply
α-Cellulose				Inert bulking agent in food
Cellulose gum	See sodium carboxymethylcellulose			
Cellulose methyl ether	See methyl cellulose			
Centaury	[*Centaurium umbellatum* Gilib.]	—	FDA §172.510	Natural flavor in alcoholic beverages only
Centaury minor herb	See centaury [*C. umbellatum* Gilib.]			
α-Cetone	See methyl α-ionone			
β-Cetone	See methyl β-ionone			
Cetone D	See methyl β-naphthyl ketone			
Cetone V	See allyl α-ionone			
Cetyl alcohol	See 1-hexadecanol			
Cetylic acid	See palmitic acid			
Ceylon cinnamon	See cinnamon [*Cinnamomum* spp.]			
Chamomile	See camomile			
Chavicyl methyl ether	See estragole			
Checkerberry, extract	See wintergreen, extract [*Gaultheria procumbens*(L.)]			
Checkerberry, oil	See wintergreen, oil [*G. procumbens* (L.)]			
Cherry bark, wild, extract	[*Prunus serotina* Ehrh.]	2276	FDA §182.1	GRAS, natural flavor extractive
Cherry laurel, leaves	[*P. laurocerasus*(L.)]		FDA §172.510	Natural flavor; not to exceed 25 ppm prussic acid
Cherry laurel, oil (ffpa)	[*P. laurocerasus*(L.)]	2277		
Cherry laurel, water	[*P. laurocerasus*(L.)]	—	FDA §172.510	
Cherry, sweet and sour, pits, extract	[*P. avium*(L.) or *P. cerasus*(L.)]	2278	FDA §172.510	Natural flavor; not to exceed 25 ppm prussic acid
Chervil	[*Anthriscus cerefolium*(L.) Hoffm.]	2279	FDA §182.1	GRAS, natural flavor and natural flavor extractive
Chestnut leaves	[*Castanea dentata*(Marsh.) Borkh.]	—	FDA §172.510	Natural flavor
Chestnut leaves, extract	[*C. dentata*(Marsh.) Borkh.]	—	FDA §172.510	
Chestnut leaves, extract solid	[*C. dentata*(Marsh.) Borkh.]	—	FDA §172.510	

REGULATORY STATUS OF DIRECT FOOD ADDITIVES (continued)

Material	Alternate Name [Source]	FEMA No.	Regulation	Limitations
Chewing-gum base	See specific natural or synthetic masticatory substance listings			
Chicle	[*Manilkara zapotilla* (Jacq.) Gilly and *M. chicle* Gilly]		FDA §121.1059	Natural (coagulated or concentrated latex) masticatory substance in chewing gum
Chicory, extract	[*Cichorium intybus* (L.)]	2280	FDA §182.1	GRAS, natural flavor extractive
Chilte	[*Cnidoscolus* (also known as *Jatropha*) *elasticus* Lundell; *C. tepiquensis* (Cost. and Gall.) McVaugh]		FDA §121.1059	Natural (coagulated or concentrated latex) masticatory substance in chewing-gum base
China bark, extract	See quillaia [*Quillaja saponaria* Molina]			
Chinese cinnamon	See cinnamon [*Cinnamomum* spp.]			
Chinoleine	See quinoline			
Chiquibul	[*Manilkara zapotilla* Gilly]		FDA §121.1059	Natural (coagulated or concentrated latex) masticatory substance in chewing-gum base
Chirata	[*Swertia chirata* (Roxb.) Buch.-Ham]	—	FDA §172.510	Natural flavor in alcoholic beverages only
Chirata, extract	[*S. chirata* Buch.-Ham.]	—	FDA §172.510	
Chives	[*Allium schoenoprasum* (L.)]		FDA §182.1	GRAS, natural flavor
Chlorine			FDA §15	Bleaching substance for flour
Chlorine dioxide	Chlorine peroxide		FDA §15	Bleaching substance for flour
Chloroform		—		
Chloropentafluoroethane			FDA §121.1065	Propellant and aerating agent for foamed or sprayed foods; product specifications apply
Chlorophyll			MID	Coloring sausage casings, oleomargarine, and shortening; in marking and branding inks for products; product specifications apply
Cholic acid	3α,7α,12α-Trihydroxy-5β-cholanic acid; cholalic acid		FDA §182.1	GRAS, emulsifying agent in dried egg whites, up to 0.1%

Choline bitartrate			FDA §182.1	GRAS, nutrient and/or dietary supplement
Choline chloride	Biocolina		FDA §182.1	GRAS, nutrient and/or dietary supplement
Chondrus extract	See Irish moss, extract			
2-Chromanone	See dihydrocoumarin			
Cinchona, bark, red	[*Cinchona succirubra* Pav. or its hybrids]	2281	FDA §172.510	Natural flavor in beverages; not more than 83 ppm total cinchona alkaloids in finished beverages
Cinchona, bark, red, extract	[*C. succirubra* Pav. or its hybrids]	2282	FDA §172.510	
Cinchona, bark, yellow	[*C. ledgeriana* Moens; *C. calisaya* Wedd.; or hybrids with other *Cinchona* spp.]	2283	FDA §172.510	Natural flavor in beverages; not more than 83 ppm total cinchona alkaloids in finished beverages
Cinchona, bark, yellow, extract	[*C. ledgeriana* Moens; *C. calisaya* Wedd.; or hybrids with other *Cinchona* spp.]	2284	FDA §172.510	
Cinchona, extract	[*C. ledgeriana* Moens and Trimen; and other *Cinchona* spp.]	2285		
Cinene	See *d*-limonene			
Cineole	See eucalyptol			
1,4-Cineole		—	FDA §172.515	Synthetic flavor/adjuvant
Cinnamal	See cinnamaldehyde			
Cinnamaldehyde	Cinnamic aldehyde	2286	FDA §182.1	GRAS, synthetic flavor/adjuvant
Cinnamaldehyde ethylene glycol acetal		2287	FDA §172.515	Synthetic flavor
Cinnamein	See benzyl cinnamate			
Cinnamic acid	β-Phenylacrylic acid	2288	FDA §172.515	Synthetic flavor
Cinnamic acid, benzyl ester	See benzyl cinnamate			
Cinnamic acid, butyl ester	See butyl cinnamate			
Cinnamic acid, ethyl ester	See ethyl cinnamate			
Cinnamic acid, isoamyl ester	See isoamyl cinnamate			
Cinnamic acid, isobutyl ester	See isobutyl cinnamate			
Cinnamic acid, tetrahydrofurfuryl ester	See tetrahydrofurfuryl cinnamate			

REGULATORY STATUS OF DIRECT FOOD ADDITIVES (continued)

Material	Alternate Name [Source]	FEMA No.	Regulation	Limitations
Cinnamic alcohol	See cinnamyl alcohol			
Cinnamic aldehyde	See cinnamaldehyde			
Cinnamic aldehyde ethylene glycol acetal	See cinnamaldehyde ethylene glycol acetal			
Cinnamon bark, Ceylon, extract	[*Cinnamomum zeylanicum* Nees]	2290	FDA §182.1	GRAS, natural flavor extractive
Cinnamon bark, Chinese, extract	[*C. cassia* Blume]	2290	FDA §182.1	GRAS, natural flavor extractive
Cinnamon bark, Saigon, extract	[*C. loureirii* Nees]	2290	FDA §182.1	GRAS, natural flavor extractive
Cinnamon bark, oil	[*Cinnamomum* spp.]	2291	FDA §182.1	Natural flavor
Cinnamon, Ceylon	[*C. zeylanicum* Nees]	2289	FDA §182.1	GRAS, natural flavor
Cinnamon, Chinese	[*C. cassia* Blume]	2289	FDA §182.1	GRAS, natural flavor
Cinnamon flowers	See cassia buds [*C. cassia* Blume]			
Cinnamon leaf, Ceylon, oil	See cinnamon leaf, oil [*Cinnamomum* spp.]			
Cinnamon leaf, Chinese, oil	See cinnamon leaf, oil [*Cinnamomum* spp.]			
Cinnamon leaf, oil	[*Cinnamomum* spp.]	2292	FDA §182.1	Natural flavor
Cinnamon leaf, Saigon, oil	See cinnamon leaf, oil [*Cinnamomum* spp.]			
Cinnamon, Saigon	[*C. loureirii* Nees]	2289	FDA §182.1	GRAS, natural flavor
Cinnamyl acetate		2293	FDA §172.515	Synthetic flavor
Cinnamyl alcohol	3-Phenyl-2-propen-1-ol	2294	FDA §172.515	Synthetic flavor
Cinnamyl anthranilate		2295	FDA §172.515	Synthetic flavor
Cinnamyl benzoate		—	FDA §172.515	Synthetic flavor
Cinnamyl butyrate		2296	FDA §172.515	Synthetic flavor
Cinnamyl cinnamate	Phenylallyl cinnamate	2298	FDA §172.515	Synthetic flavor
Cinnamyl formate		2299	FDA §172.515	Synthetic flavor
Cinnamylic acid	See cinnamic acid			
Cinnamyl isobutyrate		2297	FDA §172.515	Synthetic flavor
Cinnamyl isovalerate		2302	FDA §172.515	Synthetic flavor

Cinnamyl isovalerianate	See cinnamyl isovalerate			
Cinnamyl methanoate	See cinnamyl formate			
Cinnamyl 3-methylbunoate	See cinnamyl isovalerate			
Cinnamyl 2-methylpropanoate	See cinnamyl isobutyrate			
Cinnamyl phenylacetate		2300	FDA §172.515	Synthetic flavor
Cinnamyl β-phenylacrylate	See cinnamyl cinnamate			
Cinnamyl propionate		2301	FDA §172.515	Synthetic flavor
Cinnamyl α-toluate	See cinnamyl phenylacetate			
Cinnamic acid, allyl ester	See allyl cinnamate			
Cinncloval	See cinnamaldehyde ethylene glycol acetal			
Cire d'abeille, absolute	See beeswax, white [*Apis mellifera* (L.)]			
Citral	3,7-Dimethyl-2,6-octadien-1-al; geranial (*trans* isomer); neral (*cis* isomer)	2303	FDA §182.1	GRAS, synthetic flavor
Citral diethyl acetal	3,7-Dimethyl-2,6-octadienal diethyl acetal	2304	FDA §172.515	Synthetic flavor
Citral dimethyl acetal	3,7-Dimethyl-2,6-octadienal dimethyl acetal	2305	FDA §172.515	Synthetic flavor
Citral propylene glycol acetal		—	FDA §172.515	Synthetic flavor/adjuvant
Citric acid	2-Hydroxy-1,2,3-propanetricarboxylic acid	2306	FDA §182.1	GRAS, sequestrant
			FDA §182.1	GRAS, miscellaneous and/or general-purpose food additive
			FDA §8.303	In production of caramel
			MID	To increase effectiveness of antioxidants in lard, shortening, and unsmoked dry sausage; at 0.01% alone or with antioxidants in lard or shortening; 0.001% in unsmoked dry sausage with 0.003% BHA
			MID	To flavor oleomargarine and chili con carne
			MID	To accelerate color fixing in cured pork and beef cuts and in cured, commi-

REGULATORY STATUS OF DIRECT FOOD ADDITIVES (continued)

Material	Alternatate Name [Source]	FEMA No.	Regulation	Limitations
				nuted meat food products; may be used in cured products to replace up to 50% of ascorbic acid, erythorbic acid, sodium ascorbate, or sodium erythrobate used
			MID	To prevent clotting of fresh beef blood, at 0.2% with or without water; not more than 2 parts water to 1 part citric acid should be used
			MID	To increase effectiveness of antioxidants in frozen, fresh pork sausages and freeze-dried meats, at 0.01% in combination with antioxidants
Citridic acid	See aconitic acid			
Citroflex® A-4	See tributyl acetylcitrate			
Citronellal	3,7-Dimethyl-6-octenal; rhodinal	2307	FDA §172.515	Synthetic flavor
Citronellal hydrate	See hydroxycitronellal			
Citronella, oil	[*Cymbopogon nardus* Rendle]	2308	FDA §182.1	GRAS, natural flavor extractive
Citronellic acid	See 3,7-dimethyl-6-octenoic acid			
Citronellol	*d*-Citronellol; 3,7-dimethyl-6-octen-1-ol		FDA §172.515	Synthetic flavor
d-Citronellol	See citronellol			
dl-Citronellol		2309	FDA §172.515	Synthetic flavor
l-Citronellol	See rhodinol			
Citronellolhydrate	See hydroxycitronellol			
Citronelloxyacetaldehyde	6,10-Dimethyl-3-oxa-9-undecanal	2310	FDA §172.515	Synthetic flavor
Citronellyl acetate	3,7-Dimethyl-6-octen-1-yl acetate	2311	FDA §172.515	Synthetic flavor
Citronellyl butanoate	See citronellyl butyrate			
Citronellyl butyrate	3,7-Dimethyl-6-octen-1-yl butyrate	2312	FDA §172.515	Synthetic flavor
Citronellyl ethanoate	See citronellyl acetate			
Citronellyl formate	3,7-Dimethyl-6-octen-1-yl formate	2314	FDA §172.515	Synthetic flavor
Citronellyl isobutyrate	3,7-Dimethyl-6-octen-1-yl isobutyrate	2313	FDA §172.515	Synthetic flavor
Citronellyl methanoate	See citronellyl formate			

Citronellyl 2-methylpropanoate	See citronellyl isobutyrate			
Citronellyl pentanoate	See citronellyl valerate			
Citronellyl phenylacetate (see also rhodinyl phenylacetate	3,7-Dimethyl-6-octen-1-yl phenylacetate	2315	FDA §172.515	Synthetic flavor
Citronellyl propanoate	See citronellyl propionate			
Citronellyl propionate	3,7-Dimethyl-6-octen-1-yl propionate	2316	FDA §172.515	Synthetic flavor
Citronellyl α-toluate	See citronellyl phenylacetate			
Citronellyl valerate	3,7-Dimethyl-6-octen-1-yl valerate	2317	FDA §172.515	Synthetic flavor
Citronellyl valerianate	See citronellyl valerate			
Citrus peels, extract	[*Citrus* spp.]	2318	FDA §182.1	GRAS, natural flavor extractive
Citrus Red No. 2			FDA §8.201	Coloring skins of oranges; 2.0 ppm max, based on weight of whole fruit; product specifications apply
Civet, absolute	Zibeth; zibet; zibetum [Civet cats, *Viverra civetta* Schreber and *V. zibetha* Schreber]	2319	FDA §182.1	GRAS, natural extractive
Civet cats	See civet, absolute [Civet cats, *V. civetta* Schreber and *V. zibetha* Schreber]			
Civetone	See cycloheptadeca-9-en-1-one			
Civettone	See cycloheptadeca-9- en-1-one			
α-*trans*-Civettone	See cycloheptadeca-9-en-1-one			
Clary	Clary sage [*Salvia sclarea* (L.)]	2320	FDA §182.1	GRAS, natural flavor and natural flavor extractive
Clary, oil	[*S. sclarea* (L.)]	2321	FDA §182.1	Natural flavor
Clary sage	See clary [*S. sclarea* (L.)]			
Clary sage, oil	See clary, oil [*S. sclarea* (L.)]			
Clove bud, extract	[*Eugenia caryophyllata* Thumb. and *E. aromatica* (L.) Baill.]	2322	FDA §182.1	GRAS, natural flavor extractive
Clove bud, oil	[*E. caryophyllata* Thumb. and *E. aromatica* (L.) Baill.]	2323	FDA §182.1	Natural flavor
Clove bud, oleoresin	[*E. caryophyllata* Thunb. and *E. aromatica* (L.) Baill.]	2324	FDA §182.1	Natural flavor
Clove leaf, oil	[*E. caryophyllata* Thunb. and *E. aromatica* (L.) Bail.]	2325	FDA §182.1	GRAS, natural flavor extractive
Clover tops, red, extract, solid	[*Trifolium pratense* (L.) and other *Trifolium* spp.]	2326	FDA §182.1	GRAS, natural flavor and natural flavor extractive

REGULATORY STATUS OF DIRECT FOOD ADDITIVES (continued)

Material	Alternate Name [Source]	FEMA No.	Regulation	Limitations
Cloves	[*Eugenia caryophyllata* Thunb. and other *Eugenia* spp.]	2327	FDA §182.1	GRAS, natural flavor
Clove stem, oil	[*E. Caryophyllata* Thunb. and *E. aromatica* (L.) Baill.]	2328	FDA §182.1	Natural flavor
CMC	See sodium carboxymethylcellulose			
Cobalt 60			FDA §121.3002	Source of gamma radiation used for food processing
			FDA §121.3003	Source of gamma radiation for treatment of wheat and potatoes
Coca leaf, extract, decocainized	[*Erythroxylon coca* Lam. and other *Erythroxylon* spp.]	2329	FDA §182.1	GRAS, natural flavor extractive when decocainized
Cochineal	[*Coccus cacti* (L.)]	2330	FDA §8.317 and §8.6009	In production of aluminum or calcium-aluminum lakes for food and drug uses
			MID	Coloring sausage casing, oleomargarine, or shortening; for marking or branding ink on products; product specifications apply
Cochineal extract (see also carminic acid)	[*Dactylopius coccus* Costa (*Coccus cacti* (L.))]		FDA §8.317	Food color; product specifications apply
Coconut aldehyde	See γ-nonalactone			
Coffee	[*Coffea arabica* (L.) and other *Coffee* spp.]		FDA §182.1	GRAS, natural flavor extractive
Caffeine	See caffeine			
Cognac oil, green	Ethyl oenanthate	2331	FDA §182.1	GRAS, natural extractive
Cognac oil, white	Ethyl oenanthate	2332	FDA §182.1	GRAS, natural extractive
Cola nut	See kola nut			
Colophony	See rosin [*Pinus* spp.]			
Copaiba	[South American *Copaifera* (L.) spp.]	—	FDA §172.510	Natural flavor
Copaiba, oil	[South American *Copaifera* (L.) spp.]	—	FDA §172.510	
Copal, Manila			FDA §8.390	Diluent in inks for marking fruits and vegetables; product specifications apply

Copper gluconate			FDA §182.1	GRAS, nutrient and/or dietary supplement, up to 0.005%
Coriander	[*Coriandrum sativum* (L.)]	2333	FDA §182.1	Natural flavor
Coriander, oil	[*C. sativum* (L.)]	2334	FDA §182.1	GRAS, natural flavor and natural flavor extractive
Coriandrol	See linalool			
Cork oak	[*Quercus suber* (L.); *Q. occidentalis* Gray]	—	FDA §172.510	Natural flavor in alcoholic beverages only
Corn endosperm oil	Xanthophyll		FDA §8.322	In chicken feed to enhance yellow color of chicken skin and eggs; product specifications apply
Corn oil	[*Zea mays* (L.)]		FDA	GRAS
Corn silk	[*Z. mays* (L.)]	2335	FDA §182.1	GRAS, natural flavor extractive
Cornstarch			FDA	GRAS
Corn syrup			FDA	GRAS
			MID	To flavor sausage, hamburger, meat loaf, luncheon meat, chopped or pressed ham; may be used alone at 2% or in combination with corn syrup solids or glucose syrup, with the combination totaling 2%, calculated on a dry basis
Corn syrup solids			MID	To flavor sausage, hamburger, meat loaf, luncheon meat, chopped or pressed ham; may be used alone at 2% or with corn syrup or glucose syrup, with the combination totaling 2%, calculated on a dry basis
Corps praline	See maltol			
Cortexal	See *p*-isopropylphenylacetaldehyde			
Costmary	Sweet Mary [*Chrysanthemum balsamita* (L.)]	—	FDA §172.510	Natural flavor in alcoholic beverages only
Costus root, oil	[*Saussurea lappa* Clarke]	2336	FDA §172.510	Natural flavor
Cottonseed flour, partially defatted, cooked			FDA §121.1019	In or on food; product specifications apply
			FA §8.315	Food color; product specifications apply

REGULATORY STATUS OF DIRECT FOOD ADDITIVES (continued)

Material	Alternate Name [Source]	FEMA No.	Regulations	Limitations
Couch grass	See dog grass, extract [*Agropyron repens* (L.) Beauv.]			
Cream of tartar	Monopotassium tartrate; potassium acid tartrate; potassium bitartrate		FDA §182.1	GRAS, miscellaneous and/or general-purpose food additive and as a sequestrant
Creosol	See *p*-methoxy-4-methylphenol			
m-Cresol	3-Hydroxytoluene	3530		See *Food Technol.* February 1978
	1-Methyl-3-hydroxybezene	3530		See *Food Technol.*, February 1978
	3-Methylphenol	3530		See *Food Technol.*, February 1978
o-Cresol		3480		
p-Cresol	4-Cresol	2337	FDA §172.515	Synthetic flavor
4-Cresol	See *p*-cresol			
o-Cresol acetate	See *o*-tolyl acetate			
o-Cresyl acetate	See *o*-tolyl acetate			
p-Cresyl acetate	See *p*-tolyl acetate			
p-Cresyl dodecanoate	See *p*-tolyl laurate			
2-(*o*-, *m*-, *p*-Cresyl)-5-hydroxydioxane	See tolualdehyde glyceryl acetal			
2-(*o*-, *m*-, *p*-Cresyl)-5-hydroxymethyldioxolane	See tolualdehyde glyceryl acetal			
o-Cresylic acetate	See *o*-tolyl acetate			
p-Cresylic acetate	See *p*-tolyl acetate			
m-Cresylic acid	See *m*-cresol			
o-Cresylic acid	See *o*-cresol			
p-Cresylic acid	See *p*-cresol			
o-Cresyl isobutyrate	See *o*-tolyl isobutyrate			
p-Cresyl isobutyrate	See *p*-tolyl isobutyrate			
p-Cresyl isovalerate	See *p*-tolyl 3-methylbutyrate			
p-Cresyl laurate	See *p*-tolyl laurate			
p-Cresyl 3-methylbutanoate	See *p*-tolyl 3-methylbutyrate			

o-Cresyl methyl ether	See *o*-methylanisole			
p-Cresyl methyl ether	See *p*-methylanisole			
p-Cresyl octanoate	See *p*-tolyl octanoate			
p-Cresyl phenylacetate	See *p*-tolyl phenylacetate			
p-Cresyl *α*-toluate	See *p*-tolyl phenylacetate			
Cretan dittany	See dittany of Crete [*Origanum dictamnus* (L.)]			
Crocus, extract	See saffron, extract [*Crocus sativus* (L.)]			
Crown gum	[*Manilkara zapotilla* Gilly and *M. chicle* Gilly]		FDA §121.1059	Natural (coagulated or concentrated latex masticatory substance in chewing-gum base
Crystallose	See saccharin, sodium salt			
Cubebs	[*Piper cubeba* (L.) F.]	2338	FDA §172.510	Natural flavor
Cubebs, oil	[*P. cubeba* (L.) F.]	2339	FDA §172.510	
Cumaldehyde	See cuminaldehyde			
o-Cumenol	See 2-isopropylphenol			
Cumin	Cummin [*Cuminum cyminum* (L.)]	2340	FDA §182.1	GRAS, natural flavor and natural extractive
Cuminal	See cuminaldehyde			
Cumin alcohol	See *p*-isopropylbenzyl alcohol			
Cuminaldehyde	Cuminal; *p*-isopropyl benzaldehyde	2341	FDA §172.515	Synthetic flavor
Cumin, black	Black caraway [*Nigella sativa* (L.)]	2342	FDA §182.1	GRAS, natural flavor
Cuminic alcohol	See *p*-isopropylbenzyl alcohol			
Cuminic aldehyde	See cuminaldehyde			
Cumin, oil	[*Cuminum cyminum* (L.)]	2343	FDA §182.1	Natural flavor
Cuminol	See *p*-isopropylbenzyl alcohol			
Cuminylacetaldehyde	See 3-(*p*-isopropyl)phenyl propionaldehyde			
Cuminyl alcohol	See *p*-isopropylbenzyl alcohol			
Cumylacetaldehyde	See *p*-isopropylphenylacetaldehyde			
Cuprous iodide			FDA §182.1	GRAS, dietary supplement in table salt as source of dietary iodine, up to 0.01%
Curaçao peel, extract	Bitter orange [*Citrus aurantium* (L.)]	2344	FDA §182.1	
Curaçao peel, oil	Bitter orange [*C. aurantium* (L.)]	2345	FDA §182.1	
Currant buds, black, absolute	[*Ribes nigrum* (L.)]	2346	FDA §172.510	Natural flavor

REGULATORY STATUS OF DIRECT FOOD ADDITIVES (continued)

Material	Alternate Name [Source]	FEMA No.	Regulation	Limitations
Currant leaves, black	[*R. nigrum* (L.)]	—	FDA §172.510	
Cusparia bark	See angostura			
Cyclamal	See 2-methyl-3-(*p*-isopropylphenyl)-propionaldehyde			
Cyclamen aldehyde	See 2-methyl-3-(*p*-isopropylphenyl)-propionaldehyde			
Cyclaviol	See 2-methyl-3-(*p*-isopropylphenyl)-propionaldehyde			
α-Cyclocitrylideneacetone	See α-ionone			
β-Cyclocitrylideneacetone	See β-ionone			
α-Cyclocitrylidenebutanone	See methyl α-ionone			
β-Cyclocitrylidenebutanone	See methyl β-ionone			
α- Cyclocitrylidenemethyl butenyl ketone	See allyl α-ionone			
α-Cyclocitrylidenemethyl ethyl ketone	See methyl α-ionone			
Cyclo-1,13-ethylenedioxy-tridecan-1,13-dione	See ethyl brassylate			
Cycloheptadeca-9-en-1-one		3425		
Cycloheptadecen-9-one-1	See cycloheptadeca-9-en-1-one			
9-Cycloheptadecen-1-one	See cycloheptadeca-9-en-1-one			
Cyclohexane	Hexahydrobenzene		FDA §8.390 and §8.6200	Diluent in inks for marking gums, confectionery, and food supplements in tablet form and inks for branding pharmaceutical forms; product specifications apply
Cyclohexane acetate	See cyclohexyl acetate			
Cyclohexaneacetic acid	Cyclohexylacetic acid	2347	FDA §172.515	Synthetic flavor
1,2-Cyclohexanedione	See 2-hydroxy-2-cyclohexen-1-one			
Cyclohexane ethyl acetate	Cyclohexylethyl acetate	2348	FDA §172.515	Synthetic flavor
Cyclohexapyrazine	See 5,6,7,8-tetrahydroquinoxaline			
Cyclohexyl acetate		2349	FDA §172.515	Synthetic flavor
Cyclohexylacetic acid	See cyclohexaneacetic acid			
Cyclohexyl anthranilate		2350	FDA §172.515	Synthetic flavor
Cyclohexyl butyrate		2351	FDA §172.515	Synthetic flavor

Cyclohexyl disulfide	See dicyclohexyl disulfide			
Cyclohexylethyl acetate	See cyclohexane ethyl acetate			
Cyclohexyl formate		2353	FDA §172.515	Synthetic flavor
Cyclohexyl isopentanoate	See cyclohexyl isovalerate			
Cyclohexyl isovalerate		2355	FDA §172.515	Synthetic flavor
Cyclohexyl isovalerianate	See cyclohexyl isovalerate			
Cyclohexyl 3-methylbutanoate	See cyclohexyl isovalerate			
Cyclohexyl β-phenylacrylate	See cyclohexyl cinnamate			
Cyclohexyl 3-phenylpropenoate	See cyclohexyl cinnamate			
Cyclohexyl propionate		2354	FDA §172.510	Synthetic flavor
Cyclopentadecanolide	See ω-pentadecalactone			
Cyclopentanethiol	Cyclopentyl mercaptan	3262		See *Food Technol.*, May 1972
Cyclopentyl mercaptan	See cyclopentanethiol			
Cyclosal	See 2-methyl-3-(*p*-isopropylphenyl)-propionaldehyde			
Cyclotene	See methylcyclopentenolone			
p-Cymen-7-carboxaldehyde	See *p*-isopropylphenylacetaldehyde			
Cymene	See *p*-cymene			
p-Cymene		2356	FDA §172.510	Synthetic flavor
p-Cymene-2-ol	See carvacrol			
p-Cymen-7-ol	See *p*-isopropylbenzyl alcohol			
p-Cymen-8-ol	See *p*,α,α-trimethylbenzyl alcohol			
2-*p*-Cymenol	See carvacrol			
3-*p*-Cymenol	See thymol			
Cymol	See *p*-cymene			
Cymophenol	See carvacrol			
α-Cymophenol	See thymol			
p-Cymylpropanal	See 3-(*p*-isopropylphenyl)-propionaldehyde			
L-Cysteine		3263	FDA §17	In bakery products, up to 0.009% of flour used
			FDA §182.1	GRAS, nutrient and/or dietary supplement as free acid, hydrochloride salt, hydrate, or anhydrous form; see *Food Technol.*, May 1972, for FEMA amul-

REGULATORY STATUS OF DIRECT FOOD ADDITIVES (continued)

Material	Alternate Name [Source]	FEMA No.	Regulation	Limitations
Cystine (L- and DL-forms)			FDA §182.1	GRAS, nutrient and/or dietary supplement, as free acid, hydrochloride salt, hydrate, or anhydrous form
Dallyl di-, tri-, tetra-, and pentasulfides	See Diallyl polysulfides			
Damar gum (resin)		—	FDA §8.390	Diluent in mixtures for coloring eggshells; product specifications apply
Damascenone	See 4-(2,6,6-trimethylcyclohexa-1,3-dienyl)-but-2-en-4-one			
Damasione	See 4-(2,6,6-trimethylcyclohex-1-enyl)-but-2-en-4-one			
Damiana leaves	[*Turnera diffusa* Willd.]	—	FDA §172.510	Natural flavor
Dandelion, fluid extract	[*Taraxacum officinale* Weber and *T. laevigatum* D.C.]	2357	FDA §182.1	GRAS, natural flavor extractive
Dandelion root, extract solid	[*T. officinale* Weber and *T. laevigatum* D.C.]	2358	FDA §182.1	GRAS, natural flavor extractive
Danish agar	See furcellaran			
Davana, oil	[*Artemisia pallens* Wall.]	2359	FDA §172.510	Natural flavor
2-*trans*, 4-*trans*-Decadienal		3135		
Decahydroisopropylmethylmethylenecyclobuta(1,2:1-3,4)dicyclopentene	See β-bourbonene			
γ-Decalactone	4-Hydroxydecanoic acid, γ-lactone	2360	FDA §172.515	Synthetic flavor
δ-Decalactone	5-Hydroxydecanoic acid, δ-lactone	2361	FDA §172.515	Synthetic flavor
			FDA §172.515	In oleomargarine, alone or with δ-dodecalactone, δ-decalactone at 10 ppm max, δ-dodecalactone at 20 ppm max; product specifications apply
Decanal	Aldehyde C-10; capraldehyde; capric aldehyde; caprinaldehyde; *n*-decylaldehyde	2362	FDA §182.1	GRAS, synthetic flavor/adjuvant
Decanal dimethyl acetal	Aldehyde C-10 dimethyl acetal; capraldehyde dimethyl acetal; decylal-	2363	FDA §172.510	Synthetic flavor

	dehyde dimethyl acetal; 1,1-dimethoxydecane; 10,10-dimethoxydecane			
Decanoic acid	See capric acid			
1-Decanol	*n*-Decyl alcohol	2365	FDA §172.510	Synthetic flavor
Decanolide-1,4	See γ-decalactone			
Decanolide-1,5	See δ-decalactone			
Decanyl acetate	See decyl acetate			
2-Decenal	Decenaldehyde	2366	FDA §172.510	Synthetic flavor
4-Decenal		3264		See *Food Technol.*, May 1972
Decenaldehyde	See 2-decenal			
3-Decen-2-one	Heptylidene acetone		FDA §172.510	Synthetic flavor/adjuvant
Decyl acetate	Decanyl acetate	2367	FDA §172.510	Synthetic flavor
3-Decylacrolein	See 2-tridecenal			
n-Decyl alcohol	See 1-decanol			
Decylaldehyde (iso)	See 2,6-dimethyloctanal			
n-Decylaldehyde	See decanal			
Decylaldehyde dimethyl acetal	See decanal dimethyl acetal			
Decyl butanoate	See decyl butyrate			
Decyl butyrate		2368	FDA §172.510	Synthetic flavor
Decyl carbinol	See undecyl alcohol			
Decylenic aldehyde	See 2-decenal			
Decyl ethanoate	See decyl acetate			
Decylic acid	See decanoic acid			
Decylic alcohol	See 1-decanol			
Decyl propanoate	See decyl propionate			
Decyl propionate		239	FDA §172.510	Synthetic flavor
Defoaming agents	See specific product listings		FDA §121.1099	To inhibit foaming; product specifications apply
Dehydrated beets	See beets, dehydrated			
Dehydroacetic acid	3-Acetyl-6-methyl-2*H*-pyran-2,4(3*H*)-dione		FDA §121.1089	And/or its sodium salt, as preservative for cut or peeled squash; 65 ppm max, expressed as the acid; product specifications apply
Dehydro-β-cyclocitral	See 2,6,6-trimethylcyclohexa-1,3-dienyl methanal			

REGULATORY STATUS OF DIRECT FOOD ADDITIVES (continued)

Material	Alternate Name [Source]	FEMA No.	Regulation	Limitations
Dehydrodihydroionol		3446		
Dehydrodihydroionone		3447		
Delphinic acid	See isovaleric acid			
Deoxycholic acid	Desoxycholic acid; 3α,12α-dihydroxy-5β-cholanic acid		FDA §182.1	GRAS, emulsifying agent in dried egg whites, up to 0.1%
Dextrans			FDA §182.1	GRAS, miscellaneous and/or general-purpose food additive; average mol wt, below 100,000
Dextrin	British gum		FDA	GRAS
Dextrose	Glucose		FDA §8.303	For production of caramel
			MID	To flavor sausage, ham, and other meat products
			MID	To flavor sausage, ham, and cured meat products
Diacetyl	2,3-Butanedione	2370	FDA §182.1	GRAS, synthetic flavor/adjuvant
			MID	To flavor oleomargarine
Diacetyl tartaric acid esters			FDA §182.1	Diacetyl tartaric acid esters of mono- and diglycerides of edible fats or oils or edible fat-forming fatty acids; GRAS, emulsifiers
Diacetyl tartaric acid esters of mono- and diglycerides			MID	To emulsify rendered animal fat or its combination with vegetable fat
Diakarmon	See *d*-sorbitol			
Diallyl disulfide	See allyl disulfide			
Diallyl polysulfides	Dallyl di-, tri-, tetra-, and pentasulfides	3533		See *Food Technol.*, February 1978
	2-Propenyl polysulfides	3533		See *Food Technol.*, February 1978
Dially sulfide	See allyl sulfide			
Diallyl trisulfide		3265		See *Food Technol.*, May 1972
Diasmol	See 1,3-nonanediol acetate (mixed esters)			
Dibenzyl disulfide	See benzyl disulfide			

Dibenzyl ether	Benzyl ether	2371	FDA §172.515	Synthetic flavor
Dibenzyl ketone	See 1,3-diphenyl-2-propanone			
Dibenzyl sulfide	Benzyl sulfide		FDA §172.515	Synthetic flavor
Di-(butan-3-one-1-yl) sulfide		3335		
Dibutylbutyrolactone	See 4,4-dibutyl-γ-butyrolactone			
4,4-Dibutyl-γ-butyrolactone	4,4-Dibutyl-4-hydroxybutyric acid, γ-lactone	2372	FDA §172.515	Synthetic flavor
2,6-Di-*tert*-butyl-*p*-cresol	See butylated hydroxytoluene			
Dibutyl decanedioate	See dibutyl sebacate			
4,4-Dibutyl-4-hydroxybutyric acid, γ-lactone	See 4,4-dibutyl-γ-butyrolactone			
2,6-Di-*tert*-butyl-4-methylphenol	See butylated hydroxytoluene			
Dibutyl-1,8-octanedicarboxylate	See dibutyl sebacate			
Dibutyl sebacate	Butyl sebacate	2373	FDA §172.515	Synthetic flavor/adjuvant
Dibutyl sulfide	See butyl sulfide			
Dicalcium orthophosphate			FDA	GRAS
Dicalcium phosphate			FDA §15	Bleaching agent in flour, 6% max by weight, alone or in combination with potassium alum, calcium sulfate, magnesium carbonate, sodium aluminum sulfate, tricalcium phosphate,starch, and/or calcium carbonate, and with 1% by weight of benzoyl peroxide
			FDA §17	In bakery products, alone or in combination with calcium sulfate and/or ammonium chloride, up to 0.25% by weight of flour
Dicyclohexyl disulfide 1,1-Diethoxy-3, 7-dimethyl-2, 6-octadiene	See citral diethyl acetal	3448		
1,1-Diethoxy-3,7-dimethyl-7-octanol	See hydroxycitronellal diethyl acetal			
8,8-Diethoxy-2,6-dimethyl-2-octanol	See hydroxycitronellal diethyl acetal			
1,1-Diethoxyethane	See acetal			

REGULATORY STATUS OF DIRECT FOOD ADDITIVES (continued)

Material	Alternatate Name [Source]	FEMA No.	Regulation	Limitations
Diethyl acetal	See acetal			
Diethyl acetaldehyde	See 2-ethylbutyraldehyde			
Diethylacetic acid	See 2-ethylbutyric acid			
Diethyl anthranilate	See ethyl *n*-anthranilate			
Diethyl butanedioate	See diethyl succinate			
Diethyl decanedioate	See diethyl sebacate			
Diethyl 2,3-dihydroxybutanedioate	See diethyl tartrate			
Diethyl 2,3-dihydroxysuccinate	See diethyl tartrate			
Diethyl diketone	See 3,4-hexanedione			
1,3-Diethylacetonyl 3-methyl-3-furyl sulfide	See 3-[(2-methyl-3-furyl)-thio]-4-heptanone			
Diethylene glycol distearate			FDA §8.390	Diluent in mixtures for coloring eggshells; product specifications apply
Diethyl ethanedicarboxylate	See diethyl succinate			
Diethyl 2-hydroxybutanedioate	See diethyl malate			
Diethyl hydroxysucrinate	See diethyl malate			
Diethyl malate	Ethyl malate	2374	FDA §172.515	Synthetic flavor
Diethyl malonate	Ethyl malonate	2375	FDA §172.515	Synthetic flavor
2,3-Diethyl-5-methylpyrazine		3336		
4,5-Dihydro-3(2*H*) thiophenone		3266		See *Food Technol.*, May 1972
2,3-Dihydrotoluene	See 1-methyl-1,3-cyclohexadiene			
2,4-Dihydroxy-3-methyl-2-hexenoic acid, γ-lactone	See 5-ethyl-3-hydroxy-4-methyl-2(*H*)-furanone			
1-(*x,y*-Dihydroxyphenyl) ethanone]	See dioxyacetophenone			
1,2-Dihydroxypropane	See propylene glycol			

2,3-Dihydroxysuccinic acid	See tartaric acid			
1,1-Diisobutoxy-2-phenylethane	See phenylacetaldehyde diisobutyl acetal			
Diisobutyl carbinol	See 2,6-dimethyl-4-heptanol			
Diisobutyl ketone	See 2,6-dimethyl- 4-heptanone			
2,3-Diketobutane	See diacetyl			
Dilauryl thiodipropionate			FDA §182.1	GRAS, chemical preservative; *total content of antioxidants, not over 0.02% of fat or oil content, including essential (volatile) oil content of food*
Dill	[*Anethum graveolens* (L.)]	2382	FDA §182.1	GRAS, natural flavor and natural flavor extractive
Dill, oil	[*A. graveolens* (L.)]	2383	FDA §182.1	Natural flavor
Dill seed, Indian	[*A. sowa* Roxb. (*Peucedanum graveolens* Benth. and Hook.)]	2384	FDA §172.510	Natural flavor
Diluents	See specific product listings			
2,3-Dimercaptobutan	See 2,3-butanedithiol			
1,2-Dimercaptoethane	see 1,2-ethanedithiol			
1,6-Dimercaptohexane	See 1,6-hexanedithiol			
1,9-Dimercaptononane	See 1,9-nonanedithiol			
1,8-Dimercaptooctane	See 1,8-octanedithiol			
1,2-Dimercaptopropane	See 1,2-propanedithiol			
Diethyl 1,8-octanedicarboxylate	See diethyl sebacate			
2,3-Diethylpyrazine		3136		Not more than 10 ppm in finished food
Diethyl sebacate	Ethyl sebacate	2376	FDA §172.515	Synthetic flavor
Diethyl succinate	Ethyl succinate	2377	FDA §172.515	Synthetic flavor
Diethyl tartrate	Ethyl tartrate	2378	FDA §172.515	Synthetic flavor
2,5-Diethyltetrahydrofuran		—	FDA §172.515	Synthetic flavor
Difurfuryl disulfide	See bis(2-furfuryl) disulfide			
Difurfuryl ether		3337		
Difurfuryl sulfide	See bis(2-furfuryl) sulfide			
Dihydroanethole	See *p*-propyl anisole			
3,4-Dihydro-2*H*-1-benzopyran-2-one	See dihydrocoumarin			
2,3-Dihydrobutanedioic acid	See tartaric acid			

REGULATORY STATUS OF DIRECT FOOD ADDITIVES (continued)

Material	Alternate Name [Source]	FEMA No.	Regulation	Limitations
Dihydrocarveol	8-*p*-Menthen-2-ol; 6-methyl-3-isopropenylcyclohexanol	2379	FDA §172.515	Synthetic flavor
Dihydrocarvone		—	FDA §172.515	Synthetic flavor/adjuvant
cis-Dihydrocarvone	See *p*-Menth-8-en-2-one			
Dihydrocarvyl acetate	8-*p*-Menthen-2-yl acetate	2380	FDA §172.515	Synthetic flavor
Dihydrocinnamic acid	See 3-phenylpropionic acid			
Dihydrocinnamyl alcohol	See 3-phenyl-1-propanol			
Dihydrocitronellol	See 3,7-dimethyl-1-octanol			
Dihydrocoumarin	Hydrocoumarin	2381	FDA §172.515	Synthetic flavor
Dihydrocuminic alcohol	See *p*-mentha-1,8-dien-7-ol			
Dihydrocuminyl acetate	See perillyl acetate and *p*-mentha-1,8-dien-7-yl-acetate			
Dihydro-*p*-cymene	See α-phellandrene			
3,7-Dihydro-3,7-dimethyl-1*H*-purine-2,6-dione	See theobromine			
Dihydrogeranylacetone	See tetrahydro-pseudo-ionone			
Dihydrojasmone	See isojasmone			
5,7-Dihydro-2-methylthieno(3,4-D)pyrimidine		3338		
m-Dihydroxybenzene	See resorcinol			
Diisobutyl ketone	See 2,6-dimethyl-4-heptanone			
1,3-Diisopropylacetonyl 2-methyl-3-furyl sulfide	See 2,6-dimethyl-3-[(2-methyl-3-furyl)thio]-4-heptanone			
1,2-Dimercaptobutane	See 1,2-butanedithiol			
1,3-Dimercaptobutane	See 1,3-butanedithiol			
1,3-Dimercaptopropane	See 1,3-propanedithiol			
3,4-Dimethoxyallylbenzene	See eugenyl methyl ether			
1,2-Dimethoxy-4-allylbenzene	See eugenyl methyl ether			
1,1-Dimethoxy-2-amyl-3-phenyl-2-propene	See α-amylcinnamaldehyde dimethyl acetal			
3,4-Dimethoxybenzaldehyde	See veratraldehyde			

m-Dimethoxybenzene	Resorcinal dimethyl ether	2385	FDA §172.515	Synthetic flavor
p-Dimethoxybenzene	Dimethyl hydroquinone	2386	FDA §172.515	Synthetic flavor
1,3-Dimethoxybenzene	See *m*-dimethoxybenzene			
1,4-Dimethoxybenzene	See *p*-dimethoxybenzene			
3,4-Dimethoxybenzene carbonal	See veratraldehyde			
1,1-Dimethoxy-2-benzylideneheptane	See α-amylcinnamaldehyde dimethyl acetal			
1,1-Dimethoxydecane	See decanal dimethyl acetal			
10,10-Dimethoxydecane	See decanal dimethyl acetal			
1,1-Dimethoxy-3,7-dimethyl-2,6-octadiene	See citral dimethyl acetal			
1,1-Dimethoxy-3,7-dimethyl-7-octanol	See hydroxycitronellal dimethyl acetal			
8,8-Dimethoxy-2,6-dimethyl-2-octanol	See hydroxycitronellal dimethyl acetal			
1,1-Dimethoxyethane		3426		
1,1-Dimethoxyheptane	See heptanal dimethyl acetal			
1,1-Dimethoxyoctane	See octanal dimethyl acetal			
2,6-Dimethoxyphenol		3137		
1,1-Dimethoxy-2-phenylethane	See phenylacetaldehyde dimethyl acetal			
1,1-Dimethoxy-2-phenylpropane	See 2-phenylpropionaldehyde dimethyl acetal			
1,2-Dimethoxy-4-(2-propen-1-yl) benzene	See isoeugenyl methyl ether			
1,2-Dimethoxy-4-(1-propen-1-yl) benzene	See eugenyl methyl ether			
3,4-Dimethoxystyrene	See 3,4-dimethoxy-1-vinylbenzene			
α,α-Dimethoxytoluene	See benzaldehyde dimethyl acetal			
3,4-Dimethoxy-1-vinylbenzene		3138		
Dimethylacetal	See 1,1-dimethoxyethane			
2,4-Dimethylacetophenone	Methyl 2,4-dimethylphenyl ketone	2387	FDA §172.515	Synthetic flavor
1,4-Dimethyl-4-acetyl-1-cyclohexene		3449		
2,5-Dimethyl-3-acetylfuran	See 3-acetyl-2,5-dimethylfuran			

REGULATORY STATUS OF DIRECT FOOD ADDITIVES (continued)

Material	Alternate Name [Source]	FEMA No.	Regulations	Limitations
2,4-Dimethyl-5-acetylthiazole		3267		
2,5-Dimethyl-3-acetylthiophene	See 3-acetyl-2,5-dimethylthiophene			
3,3-Dimethylacrylic acid	See 3-methylcrotonic acid			
β,β-Dimethylacrylic acid	See 3-methylcrotonic acid			
α,α-Dimethylanisalacetone	See 1-(4-methoxyphenyl)-4-methyl-1-penten-3-one			
Dimethyl anthranilate	See methyl *N*-methylanthranilate			
2,4-Dimethylbenzaldehyde		3427		
p,α-Dimethylbenzyl alcohol	Methyl *p*-tolyl carbinol	3139		
Dimethylbenzyl carbinol	See α,α-dimethylphenethyl alcohol			
Dimethylbenzylcarbinyl acetate	See α,α-dimethylphenethyl acetate			
Dimethylbenzylcarbinyl butyrate	See α,α-dimethylphenethyl butyrate			
Dimethylbenzylcarbinyl formate	See α,α-dimethylphenethyl formate			
4,5-Dimethyl-2-benzyl-1,3-dioxolane	See phenylacetaldehyde 2,3-butylene glycol acetal			
α,α-Dimethylbenzyl isobutyrate	Phenyldimethyl carbinyl isobutyrate	2388	FDA §172.515	Synthetic flavor
α,α-Dimethylbenzyl 2-methylpropanoate	See α,α-dimethylbenzyl isobutyrate			
Dimethyl butanedioate	See dimethyl succinate			
Dimethylcarbinol	See isopropyl alcohol			
2,4-Dimethylcrotonaldehyde	See 2-methyl-2-pentenal			
1,4-Dimethylcyclohex-3-enyl methyl ketone	See 1,4-dimethyl-4-acetyl-1-cyclohexene			
3,4-Dimethyl-1,2-cyclopentadione		3268		See *Food Technol.*, May 1972

3,5-Dimethyl-1,2-cyclopentadione		3269		See *Food Technol.*, May 1972
2,3-Dimethyl-1,4-diazine	See 2,3-Dimethylpyrazine			
2,5-Dimethyl-1,4-diazine	See 2,5-dimethylpyrazine			
2,6-Dimethyl-1,4-diazine	See 2,6-dimethylpyrazine			
2,5-Dimethyl-2,5-dihydroxy-*p*-dithiane	See 2,5-dimethyl-2,5-dihydroxy-1,4-dithiane			
2,5-Dimethyl-2,5-dihydroxy-1,4-dithiane		3450		
Dimethyl disulfide	Methyl disulfide	3536	FDA §172.515	Synthetic flavor/adjuvant; see *Food Technol.*, February 1978
Dimethyl ether protocatechualdehyde	See veratraldehyde			
2,6-Dimethyl-3-ethylpyrazine	See 3-ethyl-2,6-dimethylpyrazine			
2,5-Dimethyl-3-furanthiol		3451		
(2,5-Dimethyl-3-furyl)thio-2-furoate	See 2,5-dimethyl-3-thiofuroylfuran			
(2,5-Dimethyl-3-furyl)thioisovalerate	See 2,5-dimethyl-3-thioisovalerylfuran			
Dimethylglyoxal	See diacetal			
2,6-Dimethyl-4-heptanol	Diisobutyl carbinol	3140		FEMA —flavoring, 20 ppm max in beverages, frozen desserts, candies, gelatins, puddings, and chewing gums
2,6-Dimethyl-4-heptanone	Diisobutyl ketone	3537		See *Food Technol.*, February 1978
2,6-Dimethyl-5-heptenal	Melonal	2389	FDA §172.515	Synthetic flavor
2,6-Dimethyl-2-hepten-7-al	See 2,6-dimet hyl-5-heptenal			
Dimethyl hydroquinone	See *p*-dimethoxybenzene			
1,4-Dimethyl-7-(α-hydroxyisopropyl)-δ-9,10-octahydroazulene acetate	See guaiol acetate			
3,7-Dimethyl-7-hydroxyoctanal	See hydroxycitronellal			
6,6-Dimethyl-3-hydroxy-2-methylenebicyclo[3.1.1] heptane	See 2(10)-pinen-3-0l			
5,6-Dimethyl-8-isopropenyl bicyclo-(4.4.0)-dec-1-en-3-one	See nootkatone			

REGULATORY STATUS OF DIRECT FOOD ADDITIVES (continued)

Material	Alternate Name [Source]	FEMA No.	Regulation	Limitations
1,4-Dimethyl-7-isopropenyl-δ-9,10-octahydroazulene	See guaiene			
Dimethylketol	See acetoin			
Dimethyl ketone	See acetone			
6,6-Dimethyl-2-methylene-bicyclo(3.1.1)heptane	See β-pinene			
2,6-Dimethyl-3-[(2-methyl-3-furyl)thio]-4-heptanone	3-[(2-Methyl-3-furyl)thio]-2,6-dimethyl-4-heptanone	3538		See *Food Technol.*, February 1978
	1,3-Diisopropylacetonyl 2-methyl-3-furyl sulfide	3538		See *Food Technol.*, February 1978
4-(2,2-Dimethyl-6-methylene-cyclohexyl)-3-buten-one	See γ-ionone			
4-(2,2-Dimethyl-6-methylene-cyclohexyl)-3-buten-2-one	See γ-ionone			
2,6-Dimethyl-10-methylene-2,6,11-dodecatrienal	α-Sinensal	3141		
2,2-Dimethyl-3-methylene-norbornane	See camphene			
3,3-Dimethyl-2-methylene-norcamphane	See camphene			
6,6-Dimethyl-2-methylene-norpinane	See β-pinene			
3,7-Dimethyl-2,6-octa-dienal	See citral			
3,7-Dimethyl-2,6-octa-dienal diethyl acetal	See citral diethyl acetal			
3,7-Dimethyl-2,6-octa-dienal dimethyl acetal	See citral dimethyl acetal			
2,6-Dimethyl-2,6-octadien-8-ol	See geraniol			

2,6-Dimethyl-2,7-octadien-6-ol	See linalool
3,7-Dimethyl-1,6-octadien-3-ol	See linalool
cis-2,6-Dimethyl-2,6-octadien -8-ol	See nerol
cis-3,7-Dimethyl-2,6-octadien-1-ol	See nerol
trans-3,7-Dimethyl-2,6-octadien-1-ol	See geraniol
trans-3,7-Dimethyl-2,7-octadien-1-ol	See geraniol
3,7-Dimethyl-2,6-octadien-1-thiol	See thiogeraniol
3,7-Dimethyl-1,6-octadien-3-yl acetate	See linalyl acetate
cis-3,7-Dimethyl-2,6-octadien-1-yl acetate	See neryl acetate
trans-3,7-Dimethyl-2,6-octadien-1-yl acetate	See geranyl acetate
trans-3,7-Dimethyl-2,6-octadien-1-yl acetoacetate	See geranyl acetoacetate
3,7-Dimethyl-1,6-octadien-3-yl anthranilate	See linalyl anthranilate
3,7-Dimethyl-1,6-octadien-3-yl benzoate	See linalyl benzoate
trans-3,7-Dimethyl-2,6-octadien-1-yl benzoate	See geranyl benzoate
3,7-Dimethyl-1,6-octadien-3-yl butanoate	See linalyl butyrate
cis-3,7-Dimethyl-2,6-octadien-1-yl butanoate	See neryl butyrate
trans-3,7-Dimethyl-2,6-octadien-1-yl butanoate	See geranyl butyrate
3,7-Dimethyl-1,6-octadien-3-yl butyrate	See linalyl butyrate
3,7-Dimethyl-1,6-octadien-3-yl cinnamate	See linalyl cinnamate

REGULATORY STATUS OF DIRECT FOOD ADDITIVES (continued)

Material	Alternate Name [Source]	FEMA No.	Regulation	Limitations
cis-3,7-Dimethyl-2,6-octadien-1-yl ethanoate	See neryl acetate			
trans-3,7-Dimethyl-2,6-octadien-1-yl ethanoate	See geranyl acetate			
3,7-Dimethylocta-2,6-dienyl 2-ethylbutanoate		3339		
3,7-Dimethyl-1,6-octadien-3-yl formate	See linalyl formate			
cis-3,7-Dimethyl-2,6-octadien-1-yl formate	See neryl formate			
trans-3,7-Dimethyl-2,6-octadien-1-yl formate	See geranyl formate			
3,7-Dimethyl-1,6-octadien-3-yl hexanoate	See linalyl hexanoate			
trans-3,7-Dimethyl-2,6-octadien-1-yl hexanoate	See geranyl hexanoate			
3,7-Dimethyl-1,6-octadien-3-yl isobutanoate	See linalyl isobutyrate			
cis-3,7-Dimethyl-2,6-octadien-1-yl isobutyrate	See neryl isobutyrate			
trans-3,7-Dimethyl-2,6-octadien-1-yl isobutyrate	See geranyl isobutyrate			
cis-3,7-Dimethyl-2,6-octadien-1-yl isopentanoate	See neryl isovalerate			
trans-3,7-Dimethyl-2,6-octadien-1-yl isopentanoate	See geranyl isovalerate			
3,7-Dimethyl-1,6-octadien-3-yl isovalerate	See linalyl isovalerate			
cis-3,7-Dimethyl-2,6-octadien-1-yl isovalerate	See neryl isovalerate			
trans-3,7-Dimethyl-2,6-octadien-1-yl isovalerate	See geranyl isovalerate			

cis-3,7-Dimethyl-2,6-octadien-1-yl methanoate	See neryl formate
trans-3,7-Dimethyl-2,6-octadien-1-yl methanoate	See geranyl formate
3,7-Dimethyl-1,6-octadien-3-yl 3-methylbutanoate	See linalyl isovalerate
cis-3,7-Dimethyl-2,6-octadien-1-yl 3-methylbutanoate	See neryl isovalerate
trans-3,7-Dimethyl-2,6-octadien-1-yl 3-methylbutanoate	See geranyl isovalerate
3,7-Dimethyl-1,6-octadien-3-yl 2-methylpropanoate	See linalyl isobutyrate
cis-3,7-Dimethyl-2,6-octadien-1-yl 2-methylpropanoate	See neryl isobutyrate
trans-3,7-Dimethyl-2,6-octadien-1-yl 2-methylpropanoate	See geranyl isobutyrate
3,7-Dimethyl-1,6-octadien-3-yl octanoate	See linalyl octanoate
trans-3,7-Dimethyl-2,6-octadien-1-yl 3-oxobutanoate	See geranyl acetoacetate
3,7-Dimethyl-1,6-octadien-3-yl phenylacetate	See linalyl phenylacetate
trans-3,7-Dimethyl-2,6-octadien-1-yl phenylacetate	See geranyl phenylacetate
3,7-Dimethyl-1,6-octadien-3-yl β-phenylacrylate	See linalyl cinnamate
3,7-Dimethyl-1,6-octadien-3-yl 3-phenylpropenoate	See linalyl cinnamate
3,7-Dimethyl-1,6-octadien-3-yl propanoate	See linalyl propionate
cis-3,7-Dimethyl-2,6-octadien-1-yl propanoate	See neryl propionate

REGULATORY STATUS OF DIRECT FOOD ADDITIVES (continued)

Material	Alternate Name [Source]	FEMA No.	Regulation	Limitations
trans-3,7-Dimethyl-2,6-octadien-1-yl propanoate	See geranyl propionate			
3,7-Dimethyl-1,6-octadien-3-yl propionate	See linalyl propionate			
cis-3,7-Dimethyl-2,6-octadien-1-yl propionate	See neryl propionate			
trans-3,7-Dimethyl-2,6-octadien-1-yl propionate	See geranyl propionate			
4a,5-Dimethyl-1,2,3,4,4a,5,6,7-octahydro-7-keto-3-isopropenylnaphthalene	See 4,4a,5,6,7,8-hexahydro-6-isopropenyl-4,4a-dimethyl-2(3*H*)-naphthalenone			
2,6-Dimethyloctanal	Isodecylaldehyde	2390	FDA §172.515	Synthetic flavor
3,7-Dimethyloctan-1,7-diol	See hydroxycitronellol			
3,7-Dimethyl-1,7-octanediol	See hydroxycitronellol			
3,7-Dimethyl-1-octanol	Tetrahydrogeraniol	2391	FDA §172.515	Synthetic flavor
3,7-Dimethyloctan-3-ol	See tetrahydrolinalool			
3,7-Dimethyl-1,3,6-octatriene	See ocimene *trans*-β-Ocimene	3539		See *Food Technol.*, February 1978
3,7-Dimethyl-6-octenal	See citronellal			
3,7-Dimethyl-6-octenoic acid	Citronellic acid	3142		
3,7-Dimethyl-6-octen-1-ol	See *dl*-citronellol			
3,7-Dimethyl-7-octen-1-ol	See rhodinol			
3,7-Dimethyl-(6- or 7-)-octen-1-ol	See rhodinol			
3,7-Dimethyl-6-octen-1-yl acetate	See citronellyl acetate			
3,7-Dimethyl-(6- or 7-)-octen-1-yl acetate	See rhodinyl acetate			

3,7-Dimethyl-(6- or 7-)-octen-1-yl butanoate	See rhodinyl butyrate
3,7-Dimethyl-6-octen-1-yl butyrate	See citronellyl butyrate
3,7-Dimethyl-6-octen-1-yl ethanoate	See citronellyl acetate
3,7-Dimethyl-(6- or 7-)-octen-1-yl ethanoate	See rhodinyl acetate
3,7-Dimethyl-6-octen-1-yl formate	See citronellyl formate
3,7-Dimethyl-(6- or 7-)-octen-1-yl formate	See rhodinyl formate
3,7-Dimethyl-6-octen-1-yl isobutyrate	See citronellyl isobutyrate
3,7-Dimethyl-(6- or 7-)-octen-1-yl isobutyrate	See rhodinyl isobutyrate
3,7-Dimethyl-(6- or 7-)-octen-1-yl isopentanoate	See rhodinyl isovalerate
3,7-Dimethyl-(6- or 7-)-octen-1-yl isovalerate	See rhodinyl isovalerate
3,7-Dimethyl-6-octen-1-yl methanoate	See citronellyl formate
3,7-Dimethyl-(6- or 7-)-octen-1-yl methanoate	See rhodinyl formate
3,7-Dimethyl-(6- or 7-)-octen-1-yl 3-methylbutanoate	See rhodinyl isovalerate
3,7-Dimethyl-6-octen-1-yl 2-methylpropanoate	See citronellyl isobutyrate
3,7-Dimethyl-(6- or 7-)-octen-1-yl 2-methylpropanoate	See rhodinyl isobutyrate
3,7-Dimethyl-6-octen-1-yl pentanoate	See citronellyl valerate
3,7-Dimethyl-6-octen-1-yl phenylacetate	See citronellyl phenylacetate
3,7-Dimethyl-7-octen-1-yl phenylacetate	See rhodinyl phenylacetate

REGULATORY STATUS OF DIRECT FOOD ADDITIVES (continued)

Material	Alternatate Name [Source]	FEMA No.	Regulation	Limitations
3,7-Dimethyl-6-octen-1-yl propanoate	See citronellyl propionate			
3,7-Dimethyl-(6- or 7-)-octen-1-yl propanoate	See rhodinyl propionate			
3(7-Dimethyl-6-octen-1-yl propionate	See citronellyl propionate			
3,7-Dimethyl-(6- or 7-)-octen-1-yl propionate	See rhodinyl propionate			
3,7-Dimethyl-6-octen-1-yl valerate	See citronellyl valerate			
6,10-Dimethyl-3-oxa-9-undecenal	See citronelloxyacetaldehyde			
6,6-Dimethyl-2-oxymethyl bicyclo(1.1.3)hept-2-ene	See Myrtenol			
2,5-Dimethyl paradiazine	See 2,5-dimethylpyrazine			
2,6-Dimethylparadiazine	See 2,6-dimethylpyrazine			
2,4-Dimethyl-2-pentenoic acid		3143		
α,α-Dimethylphenethanol	See α,α-dimethylphenethyl alcohol			
α,α-Dimethylphenethyl acetate	Benzyl dimethylcarbinyl acetate; benzylpropyl acetate	2392	FDA §172.515	Synthetic flavor
α,α-Dimethylphenethyl alcohol	Dimethylbenzyl carbinol	2393	FDA §172.515	Synthetic flavor
α,α-Dimethylphenethyl butyrate	Benzyl dimethylcarbinyl butyrate	2394	FDA §172.515	Synthetic flavor
Dimethylphenethyl carbinyl acetate	See 2-methyl-4-phenyl-2-butyl acetate			
Dimethylphenethyl carbinyl isobutyrate	See 2-methyl-4-phenyl-2-butyl isobutyrate			
α,α-Dimethylphenethyl formate	Benzyl dimethylcarbinyl formate	2395	§172.515	Synthetic flavor
2,5-Dimethylphenol	See 2,5-xylenol			
2,6-Dimethylphenol	See 2,6-xylenol			

3,4-Dimethylphenol	See 3,4-xylenol			
Dimethylphenylethyl carbinyl acetate	See 2-methyl-4-phenyl-2-butyl acetate			
Dimethylphenylethyl carbinyl isobutyrate	See 2-methyl-4-phenyl-2-butyl isobutyrate			
1,1-Dimethyl-3-phenylpropan-1-yl acetate	See 2-methyl-4-phenyl-2-butyl acetate			
2,5-Dimethylpiazine	See 2,5-dimethylpyrazine			
2,6-Dimethylpiazine	See 2,6-dimethylpyrazine			
Dimethyl polysiloxane	Dimethicone		FDA §121.1099	Defoaming agent, up to 10 ppm in ready-to-serve foods; 16 ppm in ready-to-serve gelatin desserts; not to be used in milk and in food for infants and invalids; product specifications apply
2,3-Dimethylpyrazine	2,3-Dimethyl-1,4-diazine	3271		See *Food Technol.*, May 1972
2,5-Dimethylpyrazine	2,5-Dimethyl-1,4-diazine	3272		See *Food Technol.*, May 1972
2,6-Dimethylpyrazine	2,6-Dimethyl-1,4-diazine	3273		See *Food Technol.*, May 1972
2,6-Dimethylpyridine	2,6-Lutidine	3540		See *Food Technol.*, February 1978
Dimethyl resorcinol	See *m*-dimethoxybenzene			
Dimethyl salicylate	See methyl *o*-methoxybenzoate			
p,α-Dimethylstyrene	*p*-Isopropenyltoluene; 1-methyl-4-isopropenylbenzene	3144		
Dimethyl succinate	Methyl succinate	2396	FDA §172.515	Synthetic flavor
Dimethyl sulfide	See methyl sulfide			
4,5-Dimethylthiazole		3274		See *Food Technol.*, May 1972
2,5-Dimethyl-3-thiofuroylfuran		3481		
2,5-Dimethyl-3-thioisovalerylfuran		3482		
Dimethyl *p*-tolyl carbinol	See *p*,α,α-trimethylbenzyl alcohol			
Dimethyl trisulfide		3275		See *Food Technol.*, May 1972
3,5-Dimethyl-1,2,4-trithiolane		—		
6,10-Dimethyl-5,9-undecadien-2-one	Geranyl acetone	3542		See *Food Technol.*, February 1978
6,10-Dimethyl-9-undecen-2-one	See tetrahydro-pseudo-ionone			

REGULATORY STATUS OF DIRECT FOOD ADDITIVES (continued)

Material	Alternate Name [Source]	FEMA No.	Regulation	Limitations
2,4-Dimethyl-5-vinylthiazole		3145		
3,7-Diethylxanthine	See thiobromine			
Dinitrogen monoxide	See nitrous oxide			
2,6-Dinitro-3-methoxy-1-methyl-4-*tert*-butylbenzene	See musk ambrette			
Dinoctyl sodium sulfosuccinate	Sulfosuccinic acid bis(2-ethylhexyl)ester *S*-sodium salt		FDA §121.1137	1. Wetting agent in fumaric acid-acidulated foods — dry gelatin dessert, dry beverage base, and fruit-juice drinks; 15 ppm max in finished gelatin dessert; 10 ppm max in finished beverage or fruit-juice drink; product specifications apply 2. In production of unrefined cane sugar; 0.5 ppm max per percentage point of sucrose in juice, syrup, or massecuite being processed; 25 ppm max in final molasses 3. Emulsifying agent for cocoa fat in noncarbonated beverages containing cocoa; 25 ppm max in finished beverage 4. Processing aid and wetting agent for fumaric acid-acidulated fruit-juice drinks, in combination with ω-hydroxypoly-(oxyethylene)poly-(oxypropylene)(53—59 mol)poly(oxyethylene)(14—16 mol)-block copolymers; 10 ppm max of DSS-block copolymer combination in finished beverage or fruit-juice drink; product specifications apply

				5. Stabilizing agent on gums and hydrophilic colloids to be used in foods as stabilizing and thickening agents; 0.5% max by weight of the gums or hydrophilic colloids; product specifications apply
				6. Dispersing agent in product conforming to §121.1229 (cocoa with dioctyl sodium sulfosuccinate for manufacturing); product specifications apply
			FDA §8.390	Diluent in mixtures for coloring egg shells; product specifications apply
			FDA §8.300(A3)	Diluent in color-additive mixtures for foods; 9 ppm max in finished food; product specifications apply
			MID	Cooling and retort water treatment agent for prevention of staining on exterior of canned goods, up to 0.05%
			MID	Hog-scale agent to remove hair from hog carcasses, when removed by subsequent cleaning operations
β-γ-Dioxopentane	See 2,3-pentanedione			
Dioxyacetophenone		—		
Dioxymethylene protocatechuic aldehyde	See piperonal			
Dipentene	*d*-Limonene		FDA §121.1059	In synthetic terpene resins in chewing-gum base; product specifications apply
Diphenyl	See biphenyl			
α,α (1)-Diphenylacetone	See 1,3-diphenyl-2-propanone			
Diphenyl disulfide	See phenyl disulfide			
1,4-Diphenyl-2,3-dithiobutane	See benzyl disulfide			
Diphenyl ketone	See benzophenone			
Diphenyl methanone	See benzophenone			
Di(phenylmethyl)disulfide	See benzyl disulfide			
1,3-Diphenyl-2-propanone	Dibenzyl ketone	2397	FDA §172.515	Synthetic flavor
Dipotassium phosphate	Potassium phosphate		FDA §19	Emulsifier in specified cheeses; alone or with other emulsifiers, up to 3% by weight of cheese
			FDA §182.1	GRAS, sequestrant

REGULATORY STATUS OF DIRECT FOOD ADDITIVES (continued)

Material	Alternate Name [Source]	FEMA No.	Regulations	Limitations
Dipropionyl	See 3,4-hexanedione			
1,3-Dipropylacetonyl 2-Methyl-3-furyl sulfide	See 4[(2-methyl 3-furyl)-thio]-5-nonanone			
Dipropyl disulfide	See propyl disulfide			
Dipropyl ketone	See 4-heptanone			
Dipropyl trisulfide		3276		See *Food Technol.*, May 1972
Disodium EDTA	Disodium ethylenediaminetetraacetate		FDA §121.1056	1. Alone, as food additive: 150 ppm max in aqueous multivitamin preparations, with iron salts as stabilizer for vitamin B_{12} 145 ppm max in canned black-eyed peas 165 ppm max in canned, cooked chick peas 165 ppm max in canned kidney beans 500 ppm max in canned strawberry pie filling 36 ppm max in coated sausage 75 ppm max in nonstandardized dressing 315 ppm max in dried banana component of ready-to-eat cereal products 75 ppm max in French dressing white potatoes, including cut potatoes 50 ppm max in gefilte fish balls or patties including liquid packing medium to inhibit discoloration 75 ppm max in mayonnaise 75 ppm max in salad dressing 100 ppm max in sandwich spread 75 ppm max in sauces

				2. In combination with calcium disodium EDTA as food additive: 75 ppm max in nonstandardized dressing 75 ppm max in French dressing 75 ppm max in mayonnaise 1000 ppm max (dry weight basis) in nonnutritive sweeteners 75 ppm max in salad dressing 100 ppm max in sandwich spread 75 ppm max in sauces. Product specifications apply
Disodium-5-guanylate		—	FDA §121.1109	Flavor enhancer in foods
Disodium-5-inosinate	Disodium salt of inosinic acid	—	FDA §121.1090	Flavoring adjuvant in foods; product specifications apply
Disodium phosphate	Sodium phosphate, dibasic	2398	FDA §18.520	In evaporated milk, up to 0.1% by weight of finished product
			FDA §16	In macaroni and noodle products; 0.5 to 1.0%
			FDA §15.140	In enriched farina; 0.5 to 1.0%
			FDA §19	Emulsifier in specified cheeses, up to 3% by weight
			FDA §182.1	GRAS, sequestrant
			MID	To decrease amount of cooked-out juices in cured hams, pork shoulder picnics, and loins, canned hams and pork shoulder picnics, chopped ham, and bacon; 5.0% of phosphate in pickle at 10% pump level; 0.5% of phosphate in product (only clear solution may be injected into product)
Disodium succinate		3277		GRAS by FEMA
2,6-Di-*tert*-butyl-4-hydroxymethyl phenol			FDA §121.1208	Antioxidant,alone or with other antioxidants, not to exceed 0.02% of fat or oil content of the food including essential oils

REGULATORY STATUS OF DIRECT FOOD ADDITIVES (continued)

Material	Alternate Name [Source]	FEMA No.	Regulation	Limitations
spiro[2,4-Dithia-1-methyl-8-oxabicyclo-[3.3.0]octane-3,3′-(1′-oxa-2′-methyl)cyclopentane]		3270		See *Food Technol.*, May 1972
spiro[2,4-Dithia-6-methyl-7-oxabicyclo-[3.3.0]octane-3,3′-(1′-oxa-2-methyl)cyclopentane]	See *spiro*[2,4-ithia-(-methyl-8-oxabicyclo-[3.3.0]octane-3,3′-(1′-oxa-2′-methyl)cyclopentane]			
2,8-Dithianon-4-en-4-carboxaldehyde		3483		
3,3′-Dithio-bis(2-methylfuran)	See Bis(2-methyl-3-furyl) disulfide			
2,2′-(Dithiodimethylene)-difuran	2-Furfuryl disulfide	3146		
2,2′-Dithiodithiophene	See 2-thienyl disulfide			
Dithioglycol	See 1,2-ethanedithiol			
Dittany of Crete	Spanish hops [Origanum dictamnus(L)]	2399	FDA §172.510	Natural flavor
Dittany root	Fraxinella [*Dictamnus albus*(L)]		FDA §172.510	Natural flavor in alcoholic beverages only
Divinyleneimine	See pyrrole			
DMB	See *p*-dimethoxybenzene			
DMBCA	See α,α-dimethylphenethyl acetate			
DMBC butyrate	See α,α-dimethylphenethyl butyrate			
DMBC formate	See α,αdimethylphenethyl formate			
D.M.P.E.C. acetate	See 2-methyl-4-phenyl-2-butyl acetate			
D.M.P.E.C. isobutyrate	See 2-methyl-4-phenyl-2-butyl isobutyrate			
D.M.P.E.C. 2-methylpropanoate	See 2-methyl-4-phenyl-2-butyl isobutyrate			
γ-Dodecalactone	4-Hydroxydodecanoic acid, γ-lactone	2400	FDA §172.515	Synthetic flavor
δ-Dodecalactone	5-Hydroxydodecanoic acid	2401	FDA §172.515	Synthetic flavor
			FDA §172.515	In oleomargarine, alone or with δ-deca-

				lacone, δ-dodecalactone at 20 ppm max, δ-decalactone at 10 ppm max
Dodecanal	See lauric aldehyde			
Dodecanoic acid	Lauric acid	2614		
1-Dodecanol	See Lauryl alcohol			
Dodecandide-1,4	See γ-dodecalactone			
Dodecanolide-1,5	See δ-dodecalactone			
Dodecanyl acetate	See lauryl acetate			
Dodecanyl ethanoate	See lauryl acetate			
2-Dodecenal	*n*-Dodecen-2-ol	202	FDA §172.515	Synthetic flavor
γ-Dodecen-6-lactone		—		
δ-Dodecen-9-lactone		—		
Dodecoic acid	See lauric acid			
Dodecyl acetate	See lauryl acetate			
Dodecyl alcohol	See lauryl alcohol			
Dodecyl isobutyrate		3452		
Dodecyl 2-methylpropanoate	See dodecyl isobutyrate			
Dog grass, extract	Quick grass; triticum [*Agropyron repens* (L.) Beauv.]	2403	FDA §182.1	GRAS, natural flavor extractive
Dracorubin, extract	See dragon's blood, extract [*Daemonorops* spp. or other botanical sources]			
Dragon's blood, extract	Dracorbin [*Daemonorops* spp. or other botanical sources]	2404	FDA §172.51	Natural flavor
Dried algae meal	See algae meal, dried			
Dried yeasts	[*Saccharomyces cerevisiae*]		FDA §17	In bakery products; 2% max by weight of flour
	[*S. cerevisiae. S. fragilis*, and dried torula yeast *Candida utilis*]		FDA §121.1125	In food; total folic acid in yeast, not to exceed 0.04 mg/g of yeast (ca. 0.008 mg of pteroylglutamic acid per g of yeast)
Dulse	[*Rhodymenia palmata* (L.)Grev.]	2405	FDA §182.1	GRAS, natural substance and natural extractive (solvent-free) used with natural flavorings
EAK	See 3-octanone			
trans-Elaidic acid	See oleic acid			

REGULATORY STATUS OF DIRECT FOOD ADDITIVES (continued)

Material	Alternate Name [Source]	FEMA No.	Regulation	Limitations
Elder flowers	[*Sambucus canadensis* (L.) or *S. nigra* (L.)]	2406	FDA §182.1	GRAS, natural flavor
Elder tree leaves	[*S. nigra* (L.)]	—	FDA §172.510	Natural flavor in alcoholic beverages only; not to exceed 25 ppm of prussic acid
Elecampane, rhizomes and roots, extract	Inula; scabwort [*Inula helenium* (L.)]	—	FDA §172.510	Natural flavor in alcoholic beverages only
Elecampane, rhizomes and roots, oil	[*I. helenium* (L.)]	—	FDA §172.510	
Elemi, gum	[*Canarium commune* (L.); *C. luzonicum* Mig.]	2407	FDA §172.510	Natural flavor
Elemi, oil	[*C. commune* (L.); *C. luzonicum* Mig.]	2408	FDA §172.510	
EMPG	See ethyl methyl phenylglycidate			
Emulsifiers and/or surface-active agents	See specific product listings		FDA §121.2541	In manufacture of articles or components of articles for use in producing, manufacturing, packing, processing, preparing, treating, packaging, transporting, or holding food; product specifications apply
Enanthal	See heptanal			
Enanthaldehyde	See heptanal			
Enanthal dimethyl acetal	See heptanal dimethyl acetal			
Enanthic acid	See heptanoic acid			
Enanthic alcohol	See heptyl alcohol			
Enanthyl alcohol	See heptyl alcohol			
Enanthylidene acetone	See 3-decen-2-one			
Enocianina	See grape skin extract			
Enzactin®	See (tri-)acetin			
Enzymes	See specific enzyme		FDA §19	In curing and developing flavor of specified cheeses
EPG	See ethyl 3-phenylglycidate			

Epichlorohydrin	*dl-α*-Epichlorohydrin		FDA §121.1031	Modifier for food starch, up to 0.3% alone or with not more than 5.0% acetic anhydride
			FDA §121.1031	Modifier for food starch, up to 0.3%, with succinic anhydride, up to 4.0%
Epoxyethane	See ethylene oxide			
1,8-Epoxy-*p*-menthane	See eucalyptol			
1,2-Epoxy-*p*-menth-4(8)-en-3-one	See piperipenone oxide			
1,6-Epoxy-*p*-menth-8-en-2-one	See *cis*-carvone oxide			
Equisetic acid	See aconitic acid			
Erigeron, oil	Fleabane [*Erigeron canadensis* (L.)]	2409	FDA §172.510	Natural flavor
Erythorbic acid	Isoascorbic acid	2410	FDA §182.1	GRAS, chemical preservative
			MID	To accelerate color fixing in cured pork and beef cuts and cured, comminuted meat food products; ¾ oz to 100 lb meat or meat byproduct; product specifications apply
d-Erythro-3-ketohexonic acid lactone	See erythorobic acid			
d-Erythro-3-oxohexonic acid lactone	See erythorbic acid			
Esdragol	See estragole			
Estragole	Esdragol; esdragon; estragon; *p*-methoxyallylbenzene	2411	FDA §172.515	Synthetic flavor
Estragole(iso)	See anethole			
Estragon, oil	Tarragon [*Artemisia dracunculus* (L.)]	2412	FDA §182.1	GRAS, natural flavor extractive
Ethanal	See acetaldehyde			
1,2-Ethanedithiol		3484		
Ethanoic acid	See acetic acid			
Ethanol	See ethyl alcohol			
Ethone	See 1-(*p*-methoxyphenyl)-1-penten-3-one			
6-Ethoxy-*m*-anol	See propenylguaethol			
p-Ethoxybenzaldehyde		2413	FDA §172.515	Synthetic flavor
2-Ethoxy-*p*-cymene	See carvacrylethyl ether			

REGULATORY STATUS OF DIRECT FOOD ADDITIVES (continued)

Material	Alternate Name [Source]	FEMA No.	Regulation	Limitations
3-Ethoxy-4-hydroxy-benzaldehyde	See ethyl vanillin			
1-Ethoxy-4-hydroxybenzene	See hydroquinane monoethyl ether			
1-Ethoxy-2-hydroxy-4-propenylbenzene	See propenylguaethol			
1-Ethoxy-2-methoxy-4-(1-propen-1-yl)-phenyl benzyl ether	See isoeugenyl ethyl ether			
o-(Ethoxymethyl)phenol		3485		
2-Ethoxynaphthalene	see *β*-naphthyl ethyl ether			
p-Ethoxyphenol	See hydroquinone monoethyl ether			
2-Ethoxy-5-propenylanisole	See isoeugenyl ethyl ether			
3-Ethoxyprotocatechualdehyde	See ethyl vanillin			
Ethoxyquin	6-Ethoxy-1,2-dihydro-2,2,4-trimethylquinoline		FDA §121.1001	Antioxidant for preservation of color in chili powder, paprika, and ground chili, 100 ppm max; product specifications apply; tolerances in animal products also apply
2-Ethoxythiazole		3340		
Ethyl abietate			Under FEMA revision	
Ethyl acetate	Acetic ether	2414	FDA §182.1	GRAS, synthetic flavor/adjuvant
			FDA §8.390	Diluent in inks for marking fruits and vegetables; product specifications apply
Ethylacetic acid	See butyric acid			
Ethyl acetoacetate	Acetoacetic ester	2415	FDA §172.515	Synthetic flavor
Ethyl acetone	See 2-pentanone			
Ethyl acetylacetate	See ethyl acetoacetate			
Ethyl 2-acetyldihydrocinnamate	See ethyl 2-acetyl-3-phenylpropionate			

Ethyl acetylformate	See ethyl pyruvate			
Ethyl *α*-acetylhydrocinnamate	See ethyl 2-acetyl-3-phenylpropionate			
Ethyl 2-acetyl-3-phenylpropionate	Ethyl benzylacetoacetate	2416	FDA §172.515	Synthetic flavor
Ethyl acetalpropanoate	See ethyl levulinate			
1-Ethyl-2-acetylpyrrole		3147		
Ethyl aconitate, mixed esters	Ethyl-2-carboxyglutaconate	2417	FDA §172.515	Synthetic flavor
3-Ethylacrolein	See 2-pentenal			
Ethyl acrylate	Acrylic acid ethyl ester	2418	FDA §172.515	Synthetic flavor
Ethyl alcohol	Alcohol, anhydrous; ethanol	2419	FDA §8.305	Ethyl alcohol, or alkaline solutions thereof, as extractant for annatto
Ethyl *n*-amyl carbinol	See 3-octanol			
Ethyl amylketone	See 3-octanone			
Ethyl *p*-anisate	Ethyl *p*-methoxybenzoate	2420	FDA §172.515	Synthetic flavor
Ethyl anthranilate	Ethyl-2-aminobenzoate	2421	FDA §172.515	Synthetic flavor
Ethylbenzene carboxylate	See ethyl benzoate			
Ethyl benzoate	Ehyl benzenecarboxylate	2422	FDA §172.515	Synthetic flavor
Ethyl 2-benzoylacetate	Benzoyl acetic ester	2423	FDA §172.515	Synthetic flavor
Ethyl benzylacetoacetate	See ethyl 2-acetyl-3-phenylpropionate			
α-Ethylbenzyl alcohol	See 1-phenyl-1-propanol			
α-Ethylbenzyl butyrate	*α*-Phenylpropyl butyrate	2424	FDA §172.515	Synthetic flavor
Ethyl 2-benzylbutyrate	See ethyl 2-ethyl-3-phenylpropanoate			
Enthylbenzyl ether	See benzyl ethyl ether			
Ethyl brassylate	Tridecanedioic acid cyclic ethylene glycol diester; cyclo-1,13-ethylenedioxytridecan-1,13-dione; ethylene glycol brassylate, cyclic diester	—	FDA §172.515	Synthetic flavor/adjuvant
2-Ethylbutanal	See 2-ethylbutyraldehyde			
Ethyl butanoate	See ethyl butyrate			
Ethyl *trans*-2-butenoate		3486	FDA §172.515	Synthetic flavor
Ethyl 2-butenoate	See ethyl crotonate			
2-Ethylbutyl acetate		2425	FDA §172.515	Synthetic flavor
β-Ethylbutyl acetate	See 2-ethylbutyl acetate			
2-Ethyl-3-butylacrolein	See 2-ethyl-2-heptenal			
Ethylbutyl carbinol	See 3-heptanol			
Ethyl *sec*-butyl ether	See *sec*-butyl ethyl ether			

REGULATORY STATUS OF DIRECT FOOD ADDITIVES (continued)

Material	Alternatate Name [Source]	FEMA No.	Regulation	Limitations
Ethylbutyl ketone	See 3-heptanone			
Ethylbutyl malonate	See butyl ethyl malonate			
2-Ethylbutyraldehyde		2426	FDA §172.515	Synthetic flavor
Ethyl butyrate		2427	FDA §182.1	GRAS, synthetic flavor and adjuvant
2-Ethylbutyric acid	α-Ethylbutyric acid	2429	FDA §172.515	
α-Ethylbutyric acid	See 2-ethylbutyric acid			
Ethyl butyrolactone	See γ-hexalactone			
γ-Ethyl-γ-butyroactone	See γ-hexalactone			
Ethyl caprate	See ethyl decanoate			
Ethyl caproate	See ethyl hexanoate			
Ethyl caprylate	See ethyl octanoate			
Ethylcarbinol	See propyl alcohol			
Ethyl 2-carboxyglutaconate	Se ethyl aconitate (mixed esters)			
Ethyl carvacrol	See carvacryl ethyl ether			
Ethyl carvacryl ether	See carvacryl ethyl ether			
Ethyl cellulose		—	FDA §121.1087	Binder and filler in dry vitamin preparations; component of protective coatings for vitamin and mineral tablets; fixative in flavoring compounds; product specifications apply
			FDA §8.390 and §8.6200	Diluent in inks for marking gum, confectionery, and food supplements in tablet form; inks for marking fruits and vegetables; inks for branding pharmaceutical forms, externally applied drugs and mixtures for coloring egg shells; product specifications apply
Ethyl cetylate	See ethyl palmitate			
Ethyl cinnamate	Ethyl phenylacrylate	2430	FDA §172.515	Synthetic flavor
Ethyl citrate	Triethyl citrate	3083		
Ethyl *p*-cresoxyacetate	See ethyl (*p*-tolyloxy)-acetate			
Ethyl crotonate	*trans*-Butenoic acid ethyl ester	—	FDA §172.515	Synthetic flavor/adjuvant

Ethyl α-crotonate	See ethyl crotoate			
Ethyl cyclohexanepropionate	Ethyl cyclohexylpropionate	2431	FDA §172.515	Synthetic flavor
Ethyl 3-cyclohexylpropanoate	See ethyl cyclohexanepropionate			
Ethyl cyclohexylpropionate	See ethyl cyclohexanepropionate			
Ethyl *trans*-2, *cis*-4-decadienoate		3148		
Ethyl decanedioate	See diethyl sebacate			
Ethyl decanoate	Ethyl caprate	2432	FDA §172.515	Synthetic flavor
Ethyl decylate	See ethyl decanoate			
2-Ethyl-1,4-diazine	See 2-ethylpyrazine			
Ethyl dihydrocinnamate	See ethyl 3-phenylpropionate			
Ethyl 2,4-diketocaproate	See ethyl 2,4-dioxohexanoate			
2-Ethyl-3, 5- or 6-dimethylpyrazine		3149		10 ppm max in finished food
3-Ethyl-2,5-dimethylpyrazine	See 2-ethyl-3,(5- or 6-)dimethylpyrazine			
3-Ethyl-2,6-dimethylpyrazine	See 2-ethyl-3,(5- or 6-)dimethylpyrazine			
3-Ethyl-2,6-dimethylpyrazine	2,6-Dimethyl-3-ethyl-pyrazine	3150		10 ppm max in finished food
Ethyl 2,4-dioxohexanoate	Ethyl 2,4-diketocaproate	3278		See *Food Technol.*, May 1972
Ethyl dodecanoate	See ethyl laurate			
Ethyl dodecylate	See ethyl laurate			
trans-1,2-Ethylenedicarboxylic acid	See fumaric acid			
Ethylene brassylate	Ethylene glycol brassylate cyclic diester	3543		See *Food Technol.*, February 1978
	Tetradecanedioic acid cyclic ethylene glycol diester			See *Food Technol.*, February 1978
trans-α,β-Ethylene-1,2-dicarboxylic acid	See fumaric acid			
Ethylene dichloride	*sym*-Dichloroethane		FDA §8.305	Solvent for annatto extract, residues 30 ppm max; total chlorinated solvent residue, 30 ppm max
Ethylene dithioglycol	See 1,2-ethanedithiol			
Ethylene glycol brassylate	See ethyl brassylate			

REGULATORY STATUS OF DIRECT FOOD ADDITIVES (continued)

Material	Alternate Name [Source]	FEMA No.	Regulation	Limitations
Ethylene glycol brassylate cyclic diester	See ethylene brassylate			
Ethylene glycol distearate			FDA §8.390	Diluent in mixtures for coloring egg shells; no penetration of color-additive mixture through eggshell into egg
Ethylene glycol monoethyl ether			FDA §8.390	Diluent in inks for marking gum, confectionery, and food supplements in tablet form; product specifications apply
Ethylene glycol monophenyl ether, isobutyrate	See 2-phenoxyethyl isobutyrate			
Ethylene mercaptan	See 1,2-ethanedithiol			
Ethylene oxide		2433		
Ethyl 2,3-epoxy-2-methyl-3-phenylpropionate	See ethyl methyl phenylglycidate			
Ethyl α,β-epoxy-β-phenylpropionate	See ethyl 3-phenylglycidate			
Ethyl ethanoate	See ethyl acetate			
Ethyl 2-ethylaminobenzoate	See ethyl *N*-ethylanthranilate			
Ethyl (2-ethylaminophenyl)methanoate	See ethyl *N*-ethylanthranilate			
Ethyl *N*-ethylanthranilate			Under FEMA review	
Ethyl 2-ethyldihydrocinnamate	See ethyl 2-ethyl-3-phenylpropanoate			
Ethyl 2-ethyl-3-phenylpropanoate		3341		
2-Ethylfenchol	See 2-ethyl-1,3,3-trimethyl-2-norbornanol			
Ethyl formate	Formic ether	2434	FDA §172.515	Synthetic flavor
Ethyl formic acid	See propionic acid			

2-Ethylfuran		—	FDA §172.515	Synthetic flavor/adjuvant
Ethyl 2-furanpropionate	Ethyl furylpropionate	2435	FDA §172.515	Synthetic flavor
Ethyl furfurylhydracrylate	See ethyl 2-furanpropionate			
2-Ethyl-3-furylacrolein	See 2-furfurylidene butyraldehyde			
Ethyl furylpropionate	See ethyl 2-furanpropionate			
4-Ethylguaiacol	4-Ethyl-2-methoxyphenol	2436	FDA §172.515	Synthetic flavor
Ethyl hendecanoate	See ethyl undecanoate			
Ethyl heptanoate	Ethyl heptoate	2437	FDA §172.515	Synthetic flavor
2-Ethyl-2-heptenal	2-Ethyl-3-butylacrolein	2438	FDA §172.515	Synthetic flavor
Ethyl heptoate	See ethyl heptanoate			
Ethyl heptylate	See ethyl heptanoate			
Ethyl hexadecanoate	Ethyl palmitate	2451	FDA §172.515	Synthetic flavor
Ethyl 2,4-hexadienoate	See ethyl sorbate			
Ethyl hexahydrophenyl propionate	See ethyl cyclohexanepropionate			
Ethyl hexanoate	Capronic ether absolute	2439	FDA §172.515	Synthetic flavor
2-Ethyl-1-hexanol	2-Ethylhexyl alcohol	3151		
Ethyl-2-hexenoate		—		
Ethyl-3-hexenoate		3342		
2-Ethyl hexylacetate	See octyl acetate			
Ethyl hexyl ketone	See 3-nonanone			
Ethyl hydrocinnamate	See ethyl 3-phenylpropionate			
Ethyl 2-hydroxybenzoate	See ethyl salicylate			
Ethyl 3-hydroxybutanoate	See ethyl 3-hydroxybutyrate			
4-Ethyl-4-hydroxybutanoic acid lactone	See γ-hexalactone			
Ethyl 3-hydroxybutyrate		3428		
Ethyl β-hydroxybutyrate	See ethyl 3-hydroxybutyrate			
3-Ethyl-2-hydroxy-2-cyclopenten-1-one		3152		
3-Ethyl-2-hydroxy-4-methylcyclopent-2-en-1-one		3453		
5-Ethyl-2-hydroxy-3-methylcyclopent-2-en-1-one		3454		
5-Ethyl-3-hydroxy-4-methyl-2(5*H*)-furanone	2,4-Dihydroxy-3-methyl-2-hexenoic acid, γ-lactone; 2-ethyl-3-methyl-4-hydroxydihydro-(2,5)-furan-5-one; 2-hydroxy-3-methyl-γ-2-hexenolactone	3153		

REGULATORY STATUS OF DIRECT FOOD ADDITIVES (continued)

Material	Alternate Name [Source]	FEMA No.	Regulation	Limitations
Ethyl 2-hydroxypropanoate	See ethyl lactate			
Ethyl α-hydroxypropionate	See ethyl lactate			
2-Ethyl-3-hydroxy-4*H*-pyran-4-one	See ethyl maltol			
Ethylidene diethyl ether	See acetal			
2-Ethylidenehexanal	See 2-butyl-2-butenal			
Ethylidene dimethyl ether	See 1,1-dimethoxyethane			
Ethyl isobutyrate		2428	FDA §172.515	Synthetic flavor
Ethyl isoeugenol	See isoeugenyl ethyl ether			
Ethyl isopentanoate	See ethyl isovalerate			
N-Ethyl-2-isopropyl-5-methylcyclohexane carboxamide	See ethyl isovalerate	3455		
Ethyl isovalerate	Ethyl β-methylbutyrate	2463	FDA §172.515	Synthetic flavor
Ethyl isovalerianate	See ethyl isovalerate			
Ethyl β-ketobutyrate	See ethyl acetoacetate			
Ethyl α-Ketopropionate	See ethyl pyruvate			
Ethyl 4-ketovalerate	See ethyl levulinate			
Ethyl lactate	Ethyl α-hydroxypropionate	2440	FDA §172.515	Synthetic flavor
Ethyl laevulate	See ethyl levulinate			
Ethyl laurate	Ethyl dodecanoate	2441	FDA §172.515	Synthetic flavor
Ethyl levulinate	Ethyl γ-ketovalerate	2442	FDA §172.515	Synthetic flavor
Ethyl malate	See diethyl malate			
Ethyl malonate	See diethyl malonate			
Ethyl maltol	2-Ethyl-3-hydroxy-4*H*-pyran-4-one	3487	FDA §172.515	Synthetic flavor/adjuvant
N-Ethyl-*p*-menthane-3-carboxamide	See *N*-ethyl-2-isopropyl-5-methylcyclohexane carboxamide			
Ethyl 2-mercaptopropionate		3279		See *Food Technol.*, May 1972
Ethyl methanedicarboxylate	See diethyl malonate			
Ethyl methanoate	See ethyl formate			

Ethyl *p*-methoxybenzoate	See ethyl *p*-anisate			
4-Ethyl-2-methoxyphenol	See 4-ethylguaiacol			
2-Ethyl-(3,5 and 6)-methoxypyrazine (85%) plus 2-methyl-(3,5 and 6)-methoxypyrazine (13%)		3280		See *Food Technol.*, May 1972
Ethyl 2-methylbutanoate	See ethyl 2-methylbutyrate			
Ethyl 3-methylbutanoate	See ethyl isovalerate			
Ethyl *trans*-2-methyl-2-butenoate	See ethyl tiglate			
Ethyl 2-methylbutyrate		2443	FDA §172.515	Synthetic flavor/adjuvant
Ethyl *trans*-2-methylcrotonate	See ethyl tiglate			
Ethyl β-methylbutyrate	See ethyl isovalerate			
3-Ethyl-4-methylcyclotene	See 3-ethyl-2-hydroxy-4-methylcyclopent-2-en-1-one			
5-Ethyl-3-methylcyclotene	See 5-ethyl-2-hydroxy-3-methylcyclopent-2-en-1-one			
2-Ethyl-3-methyl-4-hydroxydihydro-(2,5)-furan-5-one	See 5-ethyl-3-hydroxy-4-methyl-2(5*H*)-furanone	—		
Ethyl methyl ketone	See 2-butanone			
Ethyl methyl-*p*-methylphenylglycidate	See ethyl methyl-*p*-tolylglycidate			
Ethyl 2-methylpentanoate		3488		
Ethyl 2-methyl-3-pentenoate		3456		
Ethyl 2-methyl-4-pentenoate		3489		
Ethyl methyl phenylglycidate	3-Methyl-3-phenylglycidic acid ethyl ester; so-called strawberry aldehyde; aldehyde C-16	2444	FDA §182.1	GRAS, synthetic flavor/adjuvant
Ethyl 3-methyl-3-phenylglycidate	See ethyl methyl phenylglycidate			
5-Ethyl-2-methylpridine	2-methyl-5-ethylpyridine	3546		See *Food Technol.*, February 1978
Ethyl 2-methylpropanoate	See ethyl isobutyrate			
2-Ethyl-5-methylpyrazine		3154		10 ppm max in finished food
3-Ethyl-2-methylpyrazine		3155		10 ppm max in finished food

REGULATORY STATUS OF DIRECT FOOD ADDITIVES (continued)

Material	Alternate Name [Source]	FEMA No.	Regulation	Limitations
Ethyl-3-methylthiopropionate		3343		
Ethyl methyl-*p*-tolylglycidate			Under FEMA review	
Ethyl 2-methylvalerate	See ethyl 2-methylpentanoate			
Ethyl myristate	Ethyl tetradecanoate	2445	FDA §172.515	Synthetic flavor
Ethyl 2-naphthyl ether	See β-naphthyl ethyl ether			
Ethyl β-naphthyl ether	See β-naphthyl ethyl ether			
Ethyl nitrite	Nitrous ether	2446	FDA §172.515	Synthetic flavor
Ethyl nonanoate	Ethyl nonylate	2447	FDA §172.515	Synthetic flavor
Ethyl nonylate	See ethyl nonanoate			
Ethyl 2-nonynoate	Ethyl octyne carbonate	2448	FDA §172.515	Synthetic flavor
Ethyl octadecanoate		3490	FDA§172.515	Synthetic flavor
Ethyl *cis*-9-octadecenoate	See ethyl oleate			
Ethyl octanoate	Ethyl caprylate	2449	FDA §172.515	Synthetic flavor
Ethyl *cis*-4-octenoate		3344		
Ethyl octylate	See ethyl octanoate			
Ethyl octyne carbonate	See ethyl 2-nonynoate			
Ethyl oleate	Ethyl 9-octadecenoate	2450	FDA §172.515	Synthetic flavor
Ethyl 3-oxo-2-benzylbutanoate	See ethyl 2-acetyl-3-phenylpropionate			
Ethyl 3-oxobutanoate	See ethyl acetoacetate			
Ethyl 4-oxopentanoate	See ethyl levulinate			
Ethyl 2-oxopropanoate	See ethyl pyruvate			
Ethyl oxyhydrate	See rum ether			
Ethyl palmitate	See ethyl hexadecanoate			
Ethyl pelargonate	See ethyl nonanoate			
Ethyl pentanoate	See ethyl valerate			
p-Ethylphenol		3156		
Ethyl phenylacetate	α-Toluic acid	2452	FDA §172.515	Synthetic flavor
Ethyl phenylacrylate	See ethyl cinnamate			
Ethyl 4-phenylbutanoate	See ethyl 4-phenylbutyrate			

Ethyl 4-phenylbutyrate	Ethyl phenylbutyrate	2453	FDA §172.515	Synthetic flavor
Ethyl phenyl carbinol	See 1-phenyl-1-propanol			
Ethyl phenyl carbinyl butyrate	See α-ethyl benzyl butyrate			
Ethyl 3-phenyl-2,3-epoxypropionate	See ethyl 3-phenylglycidate			
Ethyl 2-phenyl-3-furoate	See 2-phenyl-3-carbethoxy furan			
Ethyl 3-phenylglycidate	Ethyl 3-phenyl-2,3-epoxypropionate	2454	FDA §172.515	Synthetic flavor
2-Ethylphenylmercaptan	See 2-ethylthiophenol			
Ethyl 3-phenyl-3-oxopropanoate	See ethyl benzoylacetate			
Ethyl 3-phenylpropenoate	See ethyl cinnamate			
Ethyl 3-phenylpropionate	Ethyl hydrocinnamate	2455	FDA §172.515	Synthetic flavor
Ethyl propanedioate	See diethyl malonate			
Ethyl propanoate	See ethyl propionate			
3-Ethyl-2-propenal	See 2-pentenal			
Ethyl 1-propene-1,2,3-tricarboxylate	See ethyl aconitate (mixed esters)			
Ethyl propenoate	See ethyl acrylate			
Ethyl propionate		2456	FDA §172.515	Synthetic flavor
Ethyl propyl carbinol	See 3-hexanol			
Ethyl propyl ketone	See 3-hexanone			
Ethyl protal	See ethyl vanillin			
Ethylprotocatechualdehyde-3-ethyl ether	See ethyl vanillin			
2-Ethylpyrazine		3281		See *Food Technol.*, May 1972
3-Ethylpyridine		3394		
β-Ethylpyridine	See 3-ethylpyridine			
2-Ethylpyromeconic acid	See ethyl maltol			
Ethyl pyrovacemate	See ethyl pyruvate			
Ethyl pyruvate	Ethyl α-ketopropionate	2457	FDA §172.515	Synthetic flavor
Ethyl salicylate	Salicylic ether	2458	FDA §172.515	Synthetic flavor
Ethyl sebacate	See diethyl sebacate			
Ethyl sorbate	Ethyl 2,4-hexadienoate	2459	FDA §172.515	Synthetic flavor
Ethyl stearate	See ethyl octadecanoate			
Ethyl succinate	See diethyl succinate			
Ethyl tartrate	See diethyl tartrate			
Ethyl tetradecanoate	See ethyl myristate			

REGULATORY STATUS OF DIRECT FOOD ADDITIVES (continued)

Material	Alternate Name [Source]	FEMA No.	Regulations	Limitations
5-Ethyl-2,3,4,5-tetramethyl-2-cyclohexen-1-one	See tetramethyl ethylcyclohexenone (mixture of isomers)			
5-Ethyl-3,4,5,6-tetramethyl-2-cyclohexen-1-one	See tetramethyl ethylcyclohexenone (mixture of isomers)			
Ethyl 2-thiazolether	See 2-ethoxythiazole			
Ethyl thioacetate		3282		See *Food Technol.*, May 1972
Ethyl thiolactate	See ethyl 2-mercaptopropionate			
2-Ethylthiophenol		3345		
Ethyl tiglate	Ethyl *trans*-2-methyl-2-butenoate	2460	FDA §172.515	Synthetic flavor
Ethyl α-toluate	See ethyl phenylacetate			
Ethyl (*p*-tolyloxy)-acetate	Ethyl *p*-cresoxyacetate	3157		
2-Ethyl-1,3,3-trimethyl-2-norbornanol		3491		
Ethyl undecanoate	Ethyl hendecanoate	3492	FDA §172.515	Synthetic flavor
Ethyl 10-undecenoate	Ethyl 10-hendecenoate	2461	FDA §172.515	Synthetic flavor
Ethyl undecylate	See ethyl undecanoate			
Ethyl undecylenate	See ethyl 10-undecenoate			
Ethyl valerate		2462	FDA §172.515	Synthetic flavor
Ethyl valerianate	See ethyl valerate			
Ethyl vanillin	3-Ethoxy-4-hydroxybenzaldehyde	2464	FDA §182.1	GRAS, synthetic flavor
Ethyl vinyl carbinol	See 1-penten-3-ol			
Ethyl vinyl ketone	See 1-penten-3-one			
Eucalyptol	Cineole; 1,8-epoxy-*p*-menthane	2465	FDA §172.515	Synthetic flavor
Eucalyptus, oil	Globulus [*Eucalyptus globulus* Labill.]	2466	FDA §172.510	Natural flavor
Eugenic acid	See eugenol			
Eugenol	4-Allyl-2-methoxyphenol	2467	FDA §182.1	GRAS, synthetic flavor
iso-Eugenol	See isoeugenol			
Eugenol acetate	See eugenyl acetate			
Eugenol benzoate	See eugenyl benzoate			
Eugenol formate	See eugenyl formate			
Eugenol methyl ether	See eugenyl methyl ether			
Eugenyl acetate	Acetyl eugenol	2469	FDA §172.515	Synthetic flavor

iso-Eugenyl acetate		2470	FDA §172.515	Synthetic flavor
Eugenyl benzoate	4-Allyl-2-methoxyphenyl benzoate	2471	FDA §172.515	Synthetic flavor
iso-Eugenyl benzyl ether	See isoeugenyl benzyl ether			
iso-Eugenyl ethyl ether	See isoeugenyl ethyl ether			
Eugenyl formate	4-Allyl-2-methoxyphenyl formate	2473	FDA §172.515	Synthetic flavor
iso-Eugenyl formate	See isoeugenyl formate			
Eugenyl methyl ether	4-Allylveratrole; methyl eugenol	2475	FDA §172.515	Synthetic flavor
iso-Eugenyl methyl ether	See isoeugenyl methyl ether			
iso-Eugenyl phenylacetate	See isoeugenyl phenylacetate			
Exaltolide	See ω-pentadecalactone			
Factitious air	See nitrous oxide			
Farnesol	3,7,11-Trimethyl-2,6,10-dodecatrien-1-ol	2478	FDA §172.515	Synthetic flavor
Farnesyl acetone	See 2,6,10-trimethyl-2,6,10-pentadecatrien-14-one			
Fatty acid esters; lactylated	See lactylated fatty acid esters of glycerol and propylene glycol			
Fatty acids	Capric, caprylic, lauric, myristic, olei, palmitic, and stearic acids and their associated fatty acids and oils		FDA §121.1070	Lubricant, binder, and defoaming agents; components in manufacture of other food-grade additives; product specifications apply
			FDA §121.1099	Compounds of defoaming agents limited to use in processing beet sugar and yeast
FD&C Colors			MID	FD&C color additives; coloring sausage casing, oleomargarine, or shortening; in marking and branding ink for products; must be certified for food use by FDA; may be mixed with approved natural colors or harmless inert materials, such as common salt or sugar
FD&C Blue No. 1	Brilliant Blue FCF e.I. 42090		FDA §8.206 and §8.501	Coloring foods (including dietary supplements) generally; product specifications apply
FD&C Blue No. 2	Indigotine e.I. 73015		FDA §8.501	Food, drug, and cosmetic use; specifications in §9.81 apply
FD&C Green No. 3	Fast Green FCF e.I. 42053		FDA §8.501	Food, drug, and cosmetic use; specifications in §9.23 apply

REGULATORY STATUS OF DIRECT FOOD ADDITIVES (continued)

Material	Alternatate Name [Source]	FEMA No.	Regulation	Limitations
FD&C Red No. 2.	Amaranth e.I. 16185			Banned in the U.S.
FD&C Red No. 3	Erythrosine e.I. 45430		FDA §8.242 and §8.501	Coloring foods generally, including dietary supplements; product specifications apply
			FDA §8.242	For food use as lake only
FD&C Red No. 4	Ponceau SX e.I. 14700			Banned in the U.S.
FD&C Red No. 40	Allura Red		FDA §8.244	Coloring foods, including dietary supplements; product specifications apply
FD&C Violet No. 1	Benzyl Violet 4B e.I. 42640			Banned in the U.S.
FD&C Yellow No. 5	Tartrazine e.I. 19140		FDA §8.275 and §8.501	Coloring foods generally; product specifications apply
			FDA §8.275	Food use of lakes only; product specifications apply
FD&C Yellow No. 6	Sunset Yellow FCF e.I. 15985		FDA §8.501	Food, drug, and cosmetic use; specifications in §9.4 apply
2-Fenchanol	See fenchyl alcohol			
D-2-Fenchanone	See *d*-fenchone			
Fenchol	See fenchyl alcohol			
d-Fenchone	*d*-1,3,3-Trimethyl-2-norbornanone	2479	FDA §172.515	Synthetic flavor
Fenchyl acetate	See 1,3,3-trimethyl-2-norbornanyl acetate			
Fenchyl alcohol	1,3,3-Trimethyl-2-orbornanol	2480	FDA §172.515	Synthetic flavor
Fennel, common	[*Foeniculum vulgare* Mill.]	2481	FDA §182.1	RAS, natural flavor
Fennel, sweet	Finocchio; Florence fennel [*F. vulgare* Mill. var. *dulce* (D.C.) Alef.]	2482	FDA §182.1	GRAS, natural flavor
Fennel, sweet, oil	[*F. vulgare* Mill. var. *dulce* (D.C.) Alef.]	2483	FDA §182.1	Natural flavor
Fenugreek	[*Trigonella foenum-graecum* (L.)]	2484	FDA §182.1	GRAS, natural flavor and natural flavor extractive
Fenugreek, extract	[*T. foenum-graecum* (L.)]	2485	FDA §182.1	Natural flavor
Fenugreek, oleoresin	[*T. foenum-graecum* (L.)]	2486	FDA §182.1	Natural flavor
Fermentation-derived, milk-clotting enzyme	[*Endothia parasitica*]		FDA §121.1199	In production of cheese; product specifications apply

Ferric phosphate		FDA §182.1	GRAS, nutrient and/or dietary supplement
Ferric pyrophosphate		FDA §182.1	GRAS, nutrient and/or dietary supplement
Ferric sodium pyrophosphate		FDA §182.1	GRAS, nutrient and/or dietary supplement
Ferrous gluconate	Fergon®	FDA §182.1	GRAS, nutrient and/or dietary supplement; *Food Chemicals Codex* specifications apply
		FDA §8.320	Coloring ripe olives
Ferrous lactate		FDA §182.1	GRAS, nutrient and/or dietary supplement
Ferrous sulfate		FDA §182.1	GRAS, nutrient and/or dietary supplement
Fincin		MID	With water, salt, and monosodium glutamate to soften tissues of beef cuts; tissue weight gain, 3% max
Finocchio	See fennel, sweet		
Fir, balsam, needles and twigs	See balsam fir, needles and twigs		
Fir ("pine"), needles and twigs	[*Abies sibirica* Ledeb.; *A. alba* Mill.; *A. sachalinensis* Masters; *A. mayriana* Miyabe and Kudo]	FDA §172.510	Natural flavor
Fir, Siberian, oil	See pine needle, oil [*Abies* spp.]		
Fleabane, oil	See erigeron, oil [*Erigeron canadensis* (L.)]		
Flomine	See α-amylcinnamaldehyde		
Florence fennel	See fennel, sweet [*Foeniculum vulgare* Mill. var. *dulce* (D.C.) Alef.]		
Flosal	See α-amylcinnamaldehyde		
Floxine	See α-amylcinnamaldehyde		
Folic acid	Pteroylglutamic acid	FDA §121.1134	Component of dietary supplements; daily ingestion, not to exceed 0.10 mg
Folione	See methyl 2-octynoate		
Food starch — modified		FDA §121.1031	1. Acid-modified, by treatment with hydrochloric acid, sulfuric acid, or both 2. Bleached by one or more of the following:

REGULATORY STATUS OF DIRECT FOOD ADDITIVES (continued)

Material	Alternate Name [Source]	FEMA No.	Regulation	Limitations
				Active oxygen obtained from hydrogen peroxide and/or peracetic acid, not to exceed 0.45% of active oxygen
				Ammonium persulfate, up to 0.075%, and sulfur dioxide, up to 0.05%
				Chlorine, as sodium hypochlorite, not to exceed 0.0082 lb of chlorine per lb dry starch
				Potassium permanganate, not to exceed 0.2% (residual manganese, calculated as Mn, not to exceed 50 ppm in food starch — modified)
				Sodium chlorite, not to exceed 0.5%
				3. Oxidized, with chlorine, as sodium hypochlorite, not to exceed 0.055 lb chlorine per lb dry starch
				4. Esterified, by treatment with one of the following:
				Acetic anhydride; acetyl groups in food starch — modified, not to exceed 2.5%
				Adipic anhydride, not to exceed 0.12%, and acetic anhydride; acetyl groups in food starch — modified, not to exceed 2.5%
				1-Octenyl succinic anhydride, not to exceed 3.0%
				1-Octenyl succinic anhydride, not to exceed 2.0%, and aluminum sulfate, not to exceed 2.0%

Phosphorus oxychloride, 0.1% max
Sodium trimetaphosphate in food starch — modified, not to exceed 0.4% calculated as phosphorus
Sodium tripolyphosphate and sodium trimetaphosphate; residual phosphate in food starch — modified, not to exceed 0.4% calculated as phosphorus
Vinyl acetate; acetyl groups in food starch — modified, not to exceed 2.5%

5. Etherified by treatment with one of the following:
   - Acrolein, 0.6% max Epichlorohydrin, 0.3% max
   - Propylene oxide, 25% max
6. Esterified and etherified by treatment with one of the following:
   - Acrolein, up to 0.3%, and vinyl acetate, up to 7.5%; acetyl groups in food starch — modified, 2.5% max
   - Epichlorohydrin, not to exceed 0.3%, and acetic anhydride; acetyl groups in food starch — modified, 2.5% max
   - Epichlorohydrin, 0.3% max, and succinic anhydride, up to 4.0%
   - Phosphorus oxychloride, 0.1% max, and propylene oxide, 10% max
7. Modified by treatment with one of the following:
   - Chlorine, as sodium hypochlorite, not to exceed 0.055 lb chlorine/lb dry starch; 0.45% of active oxygen obtained from hydrogen peroxide; and propylene oxide, 25% max
   - Sodium hydroxide, 1.0% max

REGULATORY STATUS OF DIRECT FOOD ADDITIVES (continued)

Material	Alternate Name [Source]	FEMA No.	Regulation	Limitations
				8. Food starch may be modified by a combination of acid modification and/or bleaching and any one of the other treatments prescribed, subject to any limitations prescribed
Formaldehyde			FDA §121.1099	Preservative in defoaming agents limited to use in processing beet sugar and yeast; preservative in defoaming agents containing dimethylpolysiloxane, not to exceed 1.0% of dimethylpolysiloxane content
Formic acid		2487	FDA §172.515	Synthetic flavor
Formic acid, benzyl ester	See benzyl formate			
Formic ether	See ethyl formate			
Formosa camphor	See *d*-camphor			
2-Formylbenzofuran	See 2-benzofurancarboxaldehyde			
1-Formyl-2,4-dimethylbenzene	See 2,4-dimethylbenzaldehyde			
2-Formyl-6,6-dimethylbicyclo (3.1.1)-hept-2-ene		3395		
Frankincense	See olibanum, oil [*Boswellia* spp.]			
Fraxinella	See dittany, root [*Dictamnus albus* (L.)]			
Fruit juices			FDA §8.313	Coloring foods generally; Product specifications apply
Fumaric acid	*trans*-Butenedioic acid	2488	FDA §121.1130	Fumaric acid and its calcium ferrous magnesium, potassium, and sodium salts; product specifications apply; ferrous fumarate — source of iron in foods for special dietary use
2-Furaldehyde	See furfural			
2-Furanacrolein	See furyl acrolein			

2-Furancarbinol	See furfuryl alcohol			
2-Furanmethanethiol	See furfuryl mercaptan			
2-Furanmethanethiol formate		3158		
Furcellaran	[*Furcellaria fastigiata*, family Florideae, order Gigartinales]		FDA §121.1068	Emulsifier, stabilizer, and thickener in foods; product specifications apply
Furcellaran, salts			FDA §121.1069	Emulsifier, stabilizer, and thickener in foods; product specifications apply
Furfural	2-Furaldehyde; pyromucic aldehyde	2489		
Furfuralacetone	See 4-(2-furyl)-3-buten-2-one			
Furfuralcohol	See furfuryl alcohol			
Furfuryl acetate		2490		
Furfuryl alcohol	2-Furancarbinol	2491		
Furfuryl butyrate		—		
α-Furfuryl caprylate	See α-furfuryl octanoate			
Furfuryl disulfide	See bis(2-furfuryl) disulfide			
2-Furfuryl disulfide	See 2,2′-(dithiodimethylene)-difuran			
Furfuryl ether	See difurfuryl ether			
Furfurylidene acetone	See 4-(2-furyl)-3-buten-2-one			
2-Furfurylidene butyraldehyde		2492		
Furfurylidene-2-propanal		—		
2-Furfurylidene-phenylacetaldehyde	See 2-phenyl-3-(2-furyl)-prop-2-enal			
Furfuryl isopropyl sulfide		3161		
Furfuryl isovalerate	See furfuryl 3-methylbutanoate			
Furfuryl mercaptan	2-Furanmethanethiol	2493		
Furfuryl 3-methylbutanoate	Furfuryl isovalerate	3283		See *Food Technol.*, May 1972
Furfuryl methyl ether		3159		
Furfuryl methyl sulfide		3160		
2-Furfuryl monosulfide	See 2,2′-(thiodimethylene)-difuran			
α-Furfuryl octanoate		3396		
α-Furfuryl pentanoate		3397		
Furfuryl propionate		3346		
N-Furfurylpyrrole	1-(2-Furfuryl)pyrrole	3284		See *Food Technol.*, May 1972
1-(2-Furfuryl)pyrrole	See *N*-furfurylpyrrole			
Furfuryl sulfide	See bis(2-furfuryl) sulfide			
Furfuryl thioacetate		3162		

REGULATORY STATUS OF DIRECT FOOD ADDITIVES (continued)

Material	Alternate Name [Source]	FEMA No.	Regulation	Limitations
Furfuryl thiopropionate		3347		
α-Furfuryl valerate	See α-furfuryl pentanoate			
Furoyl acetone	See 1-(2-furyl)-1,3-butanedione			
Furyl acetone	See 1-furyl-2-propanone; (2-Furyl)-2-propanone			
Furyl acrolein	2-Furanacrolein	2494		
1-(2-Furyl)-1,3-butanedione	See 4-(2-furyl)-3-buten-2-one			
4-(2-Furyl)-3-buten-2-one	Furfurylidene acetone; 1-(2-furyl)-1,3-butanedione	2495		
2-Furylcarbinol	See furfuryl alcohol			
α-Furylcarbinol	See furfuryl alcohol			
2-Furyl methyl butanoate	See furfuryl butyrate			
2-Furyl methyl ketone	2-Acetylfuran	3163		
1-Furyl-2-propanone	See (2-furyl)-2-propanone			
(2-Furyl)-2-propanone	Furyl acetone; 1-furyl-2-propanone	2496	FDA §172.515	Synthetic flavor
Fusel oil, refined	Mixed amyl alcohols	2497	FDA §172.515	Synthetic flavor
Galanga, greater	[*Alpinia galanga* Willd.]		FDA §172.510	Natural flavor in alcoholic beverages only
Galangal root	Galanga; Chinese ginger [*A. officinarum* Hance]	2498	FDA §182.1	GRAS, natural flavor and natural flavor extractive
Galangal root, extract	[*Alpinia* spp.]	2499	FDA §172.510	
Galangal root, oil	[*Alpinia* spp.]	2500	FDA §172.510	
Galanga, root	See galangal root [*A. officinarum* Hance]			
Galbanum, oil	[*Ferula galbaniflua* Boiss. and Buhse and other *Ferula* spp.]	2501	FDA §172.510	Natural flavor
Galbanum, resin	[*Ferula* spp.]	2502	FDA §172.510	
Gallotannic acid	See tannic acid [*Quercus infectoria* and related *Quercus* spp.; also in many other plants]			
Gambir	Gambir catechu; pale catechu [*Uncaria gambir* (Hunter) Roxb.]	—	FDA §172.510	Natural flavor

Gambir catechu	See catechu, black, powder [*Acacia catechu* Willd.]			
Gambir, gum	See catechu, black, powder [*Acacia catechu* Willd.]			
Garden rosemary, oil	See rosemary, oil [*Rosemarinus officinalis* (L.)]			
Garlic, oil	[*Allium sativum* (L.)]	2503	FDA §182.1	GRAS, natural flavor and natural flavor extractive
Gelatin	Puragel		FDA-GRAS	
Genet, absolute	[*Sparteum junceum* (L.)]	2504	FDA §172.510	
Genet, extract	[*S. junceum* (L.)]	2505	FDA §172.510	
Genet, flowers	[*S. junceum* (L.)]		FDA §172.510	Natural flavor
Gentian, rhizome and roots, extract	[*Gentiana lutea* (L.)]	2506	FDA §172.510	Natural flavor
Gentian, stemless	[*G. acaulis* (L.)]	—	FDA §172.510	Natural flavor in alcoholic beverages only
Geranial	See citral			
Geraniol	*trans*-3,7-Dimethyl-2,6-octadien-1-ol	2507	FDA §182.1	GRAS, synthetic flavor and adjuvant
Geraniol benzoate	See geranyl benzoate			
Geranium	[*Pelargonium* spp.]		FDA §182.1	GRAS, natural flavor and natural flavor extractive
Geranium, East Indian	See palmarosa			
Geranium, rose, oil	[*P. graveolens* L'Her.]	2508	FDA §182.1	GRAS, natural flavor extractive
Geranyl acetate	Geraniol acetate	2509	FDA §182.1	GRAS, synthetic flavor and adjuvant
Geranyl acetoacetate	*trans*-3,7-Dimethyl-2,6-octadien-1-yl acetoacetate	2510	FDA §172.515	Synthetic flavor
Geranyl acetone	See 6,10-Dimethyl-5,9-undecadien-2-one	—	FDA §172.515	Synthetic flavor; see *Food Technol.*, February 1978
Geranyl benzoate		2511	FDA §172.515	Synthetic flavor
Geranyl butyrate		2512	FDA §172.515	Synthetic flavor
Geranyl caproate	See geranyl hexanoate			
Geranyl ethanoate	See geranyl acetate			
Geranyl 2-ethylbutyrate	See 3,7-dimethylocta-2,6-dienyl 2-ethylbutanoate			
Geranyl formate		2514	FDA §172.515	Synthetic flavor
Geranyl hexanoate	Geranyl caproate	2515	FDA §172.515	Synthetic flavor
Geranyl isobutyrate		2513	FDA §172.515	Synthetic flavor
Geranyl isopentanoate	See geranyl isovalerate			

REGULATORY STATUS OF DIRECT FOOD ADDITIVES (continued)

Material	Alternate Name [Source]	FEMA No.	Regulations	Limitations
Geranyl isovalerate		2518	FDA §172.515	Synthetic flavor
Geranyl isovalerianate	See geranyl isovalerate			
Geranyl β-ketobutyrate	See geranyl acetoacetate			
Geranyl methanoate	See geranyl formate			
Geranyl 3-methylbutanoate	See geranyl isovalerate			
Geranyl 2-methylpropanoate	See geranyl isobutyrate			
Geranyl 3-oxobutanoate	See geranyl acetoacetate			
Geranyl phenylacetate	Geranyl α-toluate	2516	FDA §172.515	Synthetic flavor
Geranyl propanoate	See geranyl propionate			
Geranyl propionate	2,6-Dimethyl octadien-6-yl-8-*n*-propionate	2517	FDA §172.515	Synthetic flavor
Geranyl α-toluate	See geranyl phenylacetate			
Germander, chamaedrys	Chamaedrys [*Teucrium chamaedrys* (L.)]	—	FDA §172.510	Natural flavor in alcoholic beverages only
Germander, chamaedrys, extract	[*T. chamaedrys* (L.)]	—	FDA §172.510	
Germander, chamaedrys, extract solid	[*T. chamaedrys* (L.)]	—	FDA §172.510	
Germander, golden	[*T. polium* (L.)]		FDA §172.510	Natural flavor in alcoholic beverages only
Ghatti, gum	Indian gum [*Anogeissus latifolia* Wall.]	2519	FDA §182.1	GRAS, stabilizer
Giltex	See sodium hexametaphosphate			
Ginger	[*Zingiber officinale* Rosc.]	2520	FDA §182.1	GRAS, natural flavor and natural flavor extractive
Ginger, extract	[*Z. officinale* Rosc.]	2521	FDA §182.1	Natural flavor
Ginger, oil	[*Z. officinale* Rosc.]	2522	FDA §182.1	Natural flavor
Ginger, oleoresin	[*Z. officinale* Rosc.]	2523	FDA §182.1	Natural flavor
Ginger, wild	See snakeroot, Canadian			
Glucono-δ-lactone			MID	8 oz/100 lb of cured, comminuted meat or meat product to speed up color-fix-

				ing process and to reduce time required for smoking; must be listed in product ingredient statement
d-Glucitol	See *d*-Sorbitol			
Glucosaccharonic acid	See erythrobic acid			
Glucose pentaacetate		2524	FDA §172.515	Synthetic flavor
Glucose syrup			MID	To flavor sausage, hamburger, meat loaf, luncheon meat, chopped or pressed ham; alone at 2% or with corn-syrup solids or corn syrup, with combination totaling 2% calculated on dry basis
Glutamic acid			FDA §182.1	GRAS, salt substitute
L-Glutamic acid		3285		See *Food Technol.*, May 1972
Glutamic acid hydrochloride			FDA §182.1	GRAS, salt substitute
Gluten, gum			FDA §16	Optional use in macaroni and food products, provided protein content of food does not exceed 13% by weight
Glycerides (mono- and di-)			FDA §20	In frozen desserts, not more than 0.2% by weight of finished product
			FDA §182.1	Mono- and diglycerides of edible fats or oils or edible fat-forming fatty acids (except lauric), GRAS, emulsifiers
			MID	To emulsify rendered animal fat or combination of such fat with vegetable fat in lard and shortening; at 0.5% in oleomargarine
			FDA §8.305	Extractant for annatto; mono- and diglycerides from glycerolysis of edible vegetable fats or oils
Glycerin	Glycerol	2525	FDA §182.1	GRAS, miscellaneous and/or general-purpose food additive
Glycerin, synthetic			FDA §121.1111	Food additive; product specifications apply
Glycerol	See glycerin			
Glycerol ester of gum			FDA §121.1059	Plasticizing material in chewing-gum base

REGULATORY STATUS OF DIRECT FOOD ADDITIVES (continued)

Material	Alternate Name [Source]	FEMA No.	Regulation	Limitations
Glycerol ester of partially dimerized rosin			FDA §121.1059	Plasticizing material in chewing-gum base; product specifications apply
Glycerol ester of partially hydrogenated gum or wood rosin			FDA §121.1059	Plasticizing material in chewing-gum base
Glycerol ester of polymerized rosin			FDA §121.1059	Plasticizing material in chewing-gum base; product specifications apply
Glycerol ester of wood rosin			FDA §121.1059	Plasticizing material in chewing-gum base; product specifications apply
			FDA §121.1084	Adjusting density of citrus oils used in preparation of beverages; product specifications apply
Glycerol-lacto stearate, oleate, or palmitate			MID	To emulsify rendered animal fat or combination with vegetable fat
Glycerol tributyrate	See tributyrin			
Glyceryl-lacto esters of fatty acids			FDA §121.1004	Emulsifiers and plasticizers in food; lactic acid esters of mono- and diglycerides, manufactured from glycerin, lactic acid, and fatty acids conforming with §121.1070 and/or edible fats and oils
Glyceryl monooctadecanoate	See glyceryl monostearate			
Glyceryl monooleate	Monoolein	2526	FDA §172.515	Synthetic flavor
1-Glyceryl monooleate	See glyceryl monooleate			
Glyceryl monostearate	Glyceryl monooctadecanoate; monostearin	2527	FDA §182.1	GRAS, miscellaneous and/or general-purpose food additive
1-Glyceryl monostearate	See glyceryl monostearate			
Glyceryl triacetate	See (tri-)acetin			
Glyceryl tribenzoate		3398		
Glyceryl tributyrate	See (tri-)butyrin			
Glyceryl tripropanoate	Propionic acid, triglyceride; tripropionin	3286		See *Food Technol.*, May 1972

Glycine	Aminoacetic acid; glycocoll	3287	MID	To retard rancidity in rendered animal fat or combination of such fat with vegetable fat, up to 0.01% or up to 0.02% of combination of glycine, BHA, BHT, NDGA, and propyl gallate
			FDA §182.1	On an interim basis: GRAS, nutrient and/or dietary supplement in animal feeds, as free acid, hydrochloride salt, hydrate, or anhydrous form
			FDA §121.4002	On an interim basis: 1. Masking agent for bitter aftertaste of saccharin in carbonated, artificially sweetened beverages, not to exceed 0.2% of finished product 2. Flavor agent in butterscotch and toffee flavorings used in manufacture of frosting mixes, 0.9% max 3. Stabilizer in mono- and diglycerides prepared by glycerolysis of edible fats and oils, not to exceed 0.02% of the mono- and diglycerides; *Food Chemicals Codex* specifications apply
				See *Food Technol.*, May 1972, for FEMA max use levels
Glycocholic acid	*N*-Cholylglycine		FDA §182.1	GRAS, emulsifying agent in dried egg whites, up to 0.1%
Glycocoll	See glycine			
Glycolic acid phenyl ether	See phenoxyacetic acid			
Glycyrrhiza	See licorice root [*Glycyrrhiza glabra* (L.)]			
Glycyrrhiza, extract	See licorice, extract [*G. glabra* (L.)]			
Glycyrrhizin, ammoniated	[*G. glabra* (L.) and other *Glycyrrhiza* spp.]	2528	FDA §182.1	GRAS, natural flavor extractive
Grains of paradise	[*Aframomum melegueta* Rosc. and K. Schum.]	2529	FDA §182.1	GRAS, natural flavor

REGULATORY STATUS OF DIRECT FOOD ADDITIVES (continued)

Material	Alternate Name [Source]	FEMA No.	Regulation	Limitations
Graminis	See dog grass, extract [*Agropyron repens* (L.) Beauv.]			
Grapefruit, oil	[*Citrus paradisi* Macf.; *C. decumana* (L.)]	2530	FDA §182.1	GRAS, natural flavor extractive
Grape skin, extract	Enocianina		FDA §8.318	Coloring still and carbonated drinks and ades, beverage bases, and alcoholic beverages; product specifications apply
Green chlorophyll			MID	Coloring sausage casings, oleomargarine, and shortening; in marking or branding ink for products; may be mixed with approved synthetic dyes or harmless inert material, such as common salt or sugar
Guaiac, gum, extract	[*Guaiacum officinale* (L.); *G. sanctum* (L.)]	2531	FDA §182.1	GRAS, preservative, 0.10% in edible fats and oils
Guaiacol	*o*-Methoxyphenol	2532	FDA §172.515	Synthetic flavor
Guaiacol phenylacetate	See guaiacyl phenylacetate			
Guaiac, wood, extract	[*G. officinale* (L.)]	2533	FDA §172.510	Natural flavor
Guaiac, wood, oil	[*Guaiacum* spp.]	2534	FDA §172.510	
Guaiacyl acetate	*o*-Methoxyphenyl acetate	—	FDA §172.515	Synthetic flavor/adjuvant
Guaiacyl phenylacetate		2535	FDA §172.515	Synthetic flavor
Guaic acetate	See guaiol acetate			
Guaicwood acetate	See guaiol acetate			
Guaiene	1,4-Dimethyl-7-isopropenyl-δ-9,10-octahydroazulene	—	FDA §172.515	Synthetic flavor/adjuvant
Guai-1-en-11-ol acetate	See guaiol acetate			
Guaiol acetate	1,4-Dimethyl-7-(α-hydroxyisopropyl-δ-9,10-octahydroazulene acetate	—	FDA §172.515	Synthetic flavor/adjuvant
Guaiyl acetate	See guaiol acetate			
Guarana, gum	[*Paullinia cupana* H.B.K.]	2536	FDA §172.510	Natural flavor
Guaranine	See caffeine			

Guar, flour	See guar, gum [*Cyamopsis tetragonolobus* (L.) Taub.]			
Guar, gum	[*C. tetragonolobus* (L.) Taub.]	2537	FDA §182.1	GRAS, stabilizer
Guava	[*Psidium* spp.]		FDA §182.1	GRAS, natural flavor extractant
Gum, acacia	See acacia			
Gum, Arabic	See acacia			
Gum, Benjamin	See benzoin, resin [*Styrax* spp.]			
Gum camphor	See δ-camphor			
Gum cyamopsis	See guar, gum [*Cyamopsis tetragonolobus* (L.) Taub.]			
Gum gluten	See gluten, gum			
Gutta hang kang	[*Palaquium leiocarpum* and *P. oblongifolium* Burch.]		FDA§121.1059	Masticatory substance in chewing-gum base
Hagan phosphate	See sodium hexametaphosphate			
Haw, black, bark, extract	[*Viburnum prunifolium* (L.)]	2538	FDA §172.510	Natural flavor
H.C.A.	See α-hexylcinnamaldehyde			
Hedeona, oil	See pennyroyal, oil [*Hedeoma pulegioides* (L.) Pers.; *Mentha pulegium* (L.)]			
Heliotropine	See piperonal			
Heliotropyl acetate	See piperonyl acetate			
Heliotropyl 2-methylpropanoate	See piperonyl isobutyrate			
Helium			FDA §182.1	GRAS, miscellaneous and/or general-purpose food additive
Hemlock, needles and twigs	[*Tsuga canadensis* (L.) Carr.; *T. heterophylla* (Raf.) Sarg.]	—	FDA §172.510	Natural flavor
Hemlock, oil	See spruce, oil [*Tsuga* and *Picea* spp.]			
Hendecanal	See undecanal			
Hendecanoic acid	See undecanoic acid			
1-Hendecanol	See undecyl alcohol			
1,4-Hendecanolide	See γ-undecalactone			
2-Hendecanone	See 2-undecanone			
Hendecenal	See 10-undecenal			
10-Hendecenal	See 10-undecenal			
Hendecen-9-al	See 9-undecenal			
10-Hendecenoic acid	See 10-undecenoic acid			
10-Hendecenyl acetate	See 10-undecen-1-yl acetate			

REGULATORY STATUS OF DIRECT FOOD ADDITIVES (continued)

Material	Alternate Name [Source]	FEMA No.	Regulation	Limitations
Hendecyl methyl ketone	See 2-tridecanone			
2,4-Heptadienal		3164		
γ-Heptalactone	4-Hydroxyheptanoic acid, γ-lactone	2539	FDA §172.515	Synthetic flavor
Heptaldehyde	See heptanal			
Heptaldehyde dimethyl acetal	See heptanal dimethyl acetal			
Heptanal	Enanthaldehyde; aldehyde C-7	2540	FDA §172.515	Synthetic flavor
Heptanal dimethyl acetal		2541	FDA §172.515	Synthetic flavor
Heptanal glyceryl acetal (mixed 1,2- and 1,3-acetals)		2542	FDA §172.515	Synthetic flavor
Heptane			FDA §172.515	Synthetic flavor
2,3-Heptanedione	Acetyl valeryl	2543	FDA §172.515	Synthetic flavor
Heptanoic acid		3348		
1-Heptanol	See heptyl alcohol			
3-Heptanol	Butyl ethyl carbinol	3547		See *Food Technol.*, February 1978

2-Heptanol	Methyl *n*-amyl carbinol	3288		See *Food Technol.*, May 1972
3-Heptanol			FDA §172.515	Synthetic flavor/adjuvant
Heptanolide-1,4	See γ-heptalactone			
2-Heptanone	Methyl amyl ketone	2544	FDA §172.515	Synthetic flavor
3-Heptanone	Ethyl butyl ketone	2545	FDA §172.515	Synthetic flavor
4-Heptanone	Dipropyl ketone	2546	FDA §172.515	Synthetic flavor
2-Heptenal		3165		
trans-2-Heptenal	See 2-heptenal			
4-Heptenal (*cis* and *trans*)		3289	FDA §172.515	See *Food Technol.*, May1972
4-Heptenal diethyl acetal		3349		
2-Hepten-4-one		3399		
3-Hepten-2-one		3400		
trans-3-Heptenyl acetate		3493		
trans-3-Heptenyl isobutyrate	See *trans*-3-heptenyl 2-methylpropanoate			
trans-3-Heptenyl 2-methylpropanoate		3494		
Heptyl acetate		2547	FDA §172.515	Synthetic flavor
3-Heptylacrolein	See 2-decenal			
Heptyl alcohol	Enanthic alcohol	2548	FDA §172.515	Synthetic flavor
pri-Heptyl alcohol	See heptyl alcohol			
sec-Heptyl alcohol	See 2-heptanol			
Heptylaldehyde	See heptanal			
Heptyl butanoate	See heptyl butyrate			
Heptyl butyrate		2549	FDA §172.515	Synthetic flavor
γ-Heptyl butyrolactone	See γ-undecalactone			
γ-*n*-Heptyl-γ-butyrolactone	See γ-undecalactone			
Heptyl caprylate	See heptyl octanoate			
Heptyl carbinol	See 1-octanol			
Heptyl cinnamate		2551	FDA §172.515	Synthetic flavor

REGULATORY STATUS OF DIRECT FOOD ADDITIVES (continued)

Material	Alternatate Name [Source]	FEMA No.	Regulation	Limitations
Heptyl ethanoate	See heptyl acetate			
Heptyl formate		2552	FDA §172.515	Synthetic flavor
2-Heptylfuran		3401		
n-Heptyl *p*-hydroxybenzoate			FDA §121.1186	In fermented malt beverages to inhibit microbiological spoilage, 12 ppm max
4-*n*-Heptyl-4-hydroxybutanoic acid lactone	See γ-undecalactone			
n-Heptylic acid	See heptanoic acid			
Heptylidene acetaldehyde	See 2-nonenal			
Heptylidene acetone	See 3-decen-2-one			
Heptyl isobutyrate		2550	FDA §172.515	Synthetic flavor
Heptyl methanoate	See heptyl formate			
3-Heptyl-5-methyl-2(3*H*)-furanone		3350		
Heptyl 2-methylpropanoate	See heptyl isobutyrate			
Heptyl octanoate		2553	FDA §172.515	Synthetic flavor
Heptyl octylate	See heptyl octanoate			
Heptyl β-phenylacrylate	See heptyl cinnamate			
Heptyl 3-phenylpropenoate	See heptyl cinnamate			
α-*n*-Heptyl-γ-valerolactone	See 3-heptyl-5-methyl-2(3*H*)-furanone			
δ-*n*-Heptyl-δ-valerolactone	See δ-dodecalactone			
Hexadecanoic acid	See palmitic acid			
1-Hexadecanol	Cetyl alcohol	2554	FDA §172.515	Synthetic flavor
			FDA §8.390 and §8.6200	Diluent in inks for marking gum, confectionery, and food supplements in tablet form; inks for branding pharmaceutical forms; and ingested drugs; product specifications apply
ω-6-Hexadecenlactone	Ambrettolide 16-hydroxy-6-hexadecenoic acid, ω-lactone	2555	FDA §172.515	Synthetic flavor
6-Hexadecenolide	See ω-6-hexadecenlactone			
Hexadecylic acid	See palmitic acid			

trans, trans-2,4-Hexadienal		3429		
Hexahydrocarvacrol	See *p*-menthan-2-ol			
4,4a,5,6,7,8-Hexahydro-6-isopropenyl-4,4a-dimethyl-2(3*H*)-naphthalenone	See nootkatone			
3,4,5,8,9,10-Hexahydro-4-isopropyl-1,6-dimethylnaphthalene	See cadinene			
Hexahydrophenethyl acetate	See cyclohexancethyl acetate			
Hexahydrophenylacetic acid	See cyclohexaneacetic acid			
Hexahydropyridine	See piperidine			
Hexahydrothymol	See menthol			
γ-Hexalactone	4-Hydroxyhexanoic acid, γ-lactone; tonkalide	2556	FDA §172.515	Synthetic flavor
δ-Hexalactone		3167		
Hexaldehyde	See hexanal			
Hexamethylene dimercaptan	See 1,6-hexanedithiol			
Hexanal	Aldehyde C-6; caproic aldehyde	2557	FDA §172.515	Synthetic flavor/adjuvant
2-Hexanal			FDA §172.515	Synthetic flavor/adjuvant
Hexane			FDA §8.305	Solvent for annatto extract; residues, 25 ppm max
Hexanedioic acid	See adipic acid			
2,3-Hexanedione	Acetyl butyryl	2558	FDA §172.515	Synthetic flavor
3,4-Hexanedione		3168		
1,6-Hexanedithiol		3495		
Hexanoic acid	Caproic acid	2559	FDA §172.515	Synthetic flavor
1-Hexanol	See hexyl alcohol			
3-Hexanol		3351		
Hexanolide-1,4	See γ-hexalactone			
3-Hexanone	Ethyl propyl ketone	3290		See *Food Technol.*, May 1972
Hexanoylethanol	See 3-octanon-1-ol			
2-Hexanoylfuran	See pentyl-2-furyl ketone			
2-Hexenal		2560	FDA §172.515	Synthetic flavor
cis-3-Hexenal		2561		

REGULATORY STATUS OF DIRECT FOOD ADDITIVES (continued)

Material	Alternate Name [Source]	FEMA No.	Regulations	Limitations
cis-4-Hexenal		3496		
4-Hexene-3-one		3352		
3-Hexenoic acid		3170		
trans-2-Hexenoic acid		3169		
2-Hexen-1-ol		2562	FDA §172.515	Synthetic flavor
3-Hexen-1-ol	Leaf alcohol; Blätteralkohol	2563	FDA §172.515	Synthetic flavor
4-Hexen-1-ol		3430		
2-Hexen-4-one	See 4-hexene-3-one			
2-Hexen-1-yl acetate		2564	FDA §172.515	Synthetic flavor
cis-3-Hexen-1-yl acetate		3171		
cis-3-Hexenyl butanoate	See *cis*-3-hexenyl butyrate			
cis-3-Hexenyl butyrate		3402		
β,γ-Hexenyl *n*-butyrate	See *cis*-3-hexenyl butyrate			
cis-3-Hexenyl caproate	See *cis*-3-hexenyl hexanoate			
cis-3-Hexenyl ethanoate	See *cis*-3-hexen-1-yl acetate			
3-Hexenyl formate		3353		
cis-3-Hexenyl hexanoate		3403		
3-Hexenyl isopentanoate	See 3-hexenyl isovalerate			
3-Hexenyl isovalerate	3-Hexenyl isopentanoate; 3-hexenyl 3-methylbutanoate	3489	FDA §172.515	Synthetic flavor/adjuvant
3-Hexenyl isovalerianate	See 3-hexenyl isovalerate			
3-Hexenyl methanoate	See 3-hexenyl formate			
3-Hexenyl 2-methylbutanoate	3-hexenyl 2-methylbutyrate	3497	FDA §172.515	Synthetic flavor
3-Hexenyl 3-methylbutanoate		3498	FDA §172.515	Synthetic flavor
3-Hexenyl 3-methylbutanoate	See 3-hexenyl isovalerate			
3-Hexenyl 2-methylbutyrate	See 3-hexenyl 2-methylbutanoate			
3-Hexenyl phenylacetate		—	FDA §172.515	Synthetic flavor
cis-3-Hexenyl phenylacetate	See 3-hexenyl phenylacetate			
3-Hexenyl α-toluate	See 3-hexenyl phenylacetate			

β, γ-Hexenyl α-toluate	See 3-hexenyl phenylacetate			
Hexoic acid	See hexanoic acid			
Hexoic aldehyde	See hexanal			
Hexone	See 4-methyl-2-pentanone			
Hexyl acetate		2565	FDA §172.515	Synthetic flavor
2-Hexyl-4-acetoxytetrahydrofuran		—	FDA§172.515	Synthetic flavor
3-Hexylacrolein	See 2-nonenal			
Hexyl-alcohol	1-Hexanol	2567	FDA §172.515	Synthetic flavor
Hexyl benzoate		—		
Hexyl butanoate	See hexyl butyrate			
n-Hexyl 2-butenoate		3354		
Hexyl butyrate		2568	FDA §172.515	Synthetic flavor
γ-*n*-Hexyl-γ-butyrolactone	See γ-decalactone			
Hexyl caproate	See hexyl hexanoate			
Hexyl caprylate	See hexyl octanoate			
Hexyl carbinol	See heptyl alcohol			
α-Hexylcinnamaldehyde		2569	FDA §172.515	Synthetic flavor
α-*n*-Hexylcinnamic aldehyde	See α-hexylcinnamaldehyde			
Hexyl crotonate	See *n*-hexyl 2-butenoate			
2-Hexyl-2-cyclopenten-1-one and 2-hexylidene-cyclopentanone mixture	See isojasmone			
Hexylene glycol diacetate	See 1,3-nonanediol acetate (mixed esters)			
Hexyl ethanoate	See hexyl acetate			
Hexyl formate		2570	FDA §172.515	Synthetic flavor
Hexyl furoate	See hexyl 2-furoate			
Hexyl 2-furoate	Hexyl furoate	2571		
Hexyl hexanoate		2572	FDA §172.515	Synthetic flavor
4-*n*-Hexyl-4-hydroxy butanoic acid lactone	See γ-decalactone			
2-Hexyl-4-hydroxy-1,3-dioxane	See heptanal glyceryl acetal (mixed 1,2- and 1,3-acetals)			
(1- or 2-Hexyl-2-hydroxyethoxy) acetic acid, δ-lactone	See 2-hexyl-5- or 6-keto-1,4-dioxane			

REGULATORY STATUS OF DIRECT FOOD ADDITIVES (continued)

Material	Alternatate Name [Source]	FEMA No.	Regulation	Limitations
2-Hexyl-4-hydroxymethyl-1,3-dioxolane	See heptanal glyceryl acetal (mixed 1,2-and 1,3-acetals)			
5-*n*-Hexyl-5-hydroxypentanoic acid lactone	See 5-hydroxyundecanoic acid lactone			
2-Hexylidene cyclopentanone		2573	FDA §172.515	Synthetic flavor/adjuvant
a-Hexylidene cyclopentanone	See 2-hexylidene cyclopentanone			
2-Hexylidene cyclopentanone and 2-hexyl-2-cyclopenten-1-one (mixture)	See iso-jasmone			
Hexyl isobutyrate		3172		
Hexyl isopentanoate	See hexyl 3-methylbutanoate			
Hexyl isopentanoate	See hexyl isovalerate			
Hexyl isovalerate		—	FDA §172.515	Synthetic flavor/adjuvant
Hexyl isovalerate	See hexyl 3-methylbutanoate			
Hexyl isovalerianate	See hexyl isovalerate			
2-Hexyl-5- or 6-keto-1,4-dioxane		2574		
Hexyl methanoate	See hexyl formate			
Hexyl 2- methylbutanoate		3499	FDA §172.515	Synthetic flavor
Hexyl 2-methylbutanoate	See hexyl 2-methylbutyrate			
Hexyl 3-methylbutanoate		3500	FDA §172.515	Synthetic flavor
Hexyl 3-methylbutanoate	See hexyl isovalerate			
Hexyl 2-methylbutyrate			FDA §172.515	Synthetic flavor/adjuvant
Hexyl 2-methylbutyrate	See hexyl 2-methylbutanoate			
Hexyl methyl carbinol	See 2-octanol			
Hexyl methyl ketone	See 2-octanone			
Hexyl 2-methylpropanoate	See hexyl isobutyrate			
Hexyl octanoate		2575	FDA §172.515	Synthetic flavor
Hexyl octylate	See hexyl octanoate			
Hexyl phenylacetate		3457		

α-*n*-Hexyl-β-phenylacrolein	See α-hexylcinnamaldehyde			
3-Hexyl-1,3-propanediol acetate (mixed esters)	See 1,3-nonanediol acetate (mixed esters)			
Hexyl propanoate	See hexyl propionate			
3-Hexyl-2-propenal	See 2-nonenal			
Hexyl propionate		2576	FDA §172.515	Synthetic flavor
Hexyl α-toluate	See hexyl phenylacetate			
δ-*n*-Hexyl-δ-valerolactone	See 5-hydroxyundecanoic acid lactone			
Hickory, bark, extract	[*Carya* spp.]	2577	FDA §182.1	GRAS, natural flavor extractive
Hipberries, extract	See rose hips, extract [*Rosa* spp.]			
Histidine (L- and DL-forms)			FDA §182.1	GRAS, nutrient and/or dietary supplement, as the free acid, hydrochloride salt, hydrate, or anhydrous form
Hoarhound, extract	See horehound, extract [*Marrubium vulgare* Tourn. (L.)]			
Holy thistle	See thistle, blessed [*Cnicus benedictus* (L.)]			
Holy thistle, extract	See thistle, blessed, extract [*C. benedictus* (L.)]			
Holy thistle, extract solid	See thistle, blessed, extract solid [*C. benedictus* (L.)]			
Holy thistle, oil	See thistle, blessed, oil [*C. benedictus* (L.)]			
Homoanisaldehyde	See *p*-ethoxybenzaldehyde			
Homocatechol monomethyl ether	See 2-methoxy-4-methylphenol			
Homocreosol	See 4-ethylguaiacol			
β-Homocyclocitral	See 2,6,6-trimethyl-1-cyclohexen-1-acetaldehyde			
Hops, extract	[*Humulus lupulus* (L.)]	2578	FDA §182.1	GRAS, natural flavor extractive
Hops, extract, modified			FDA §121.1082	Flavoring agent in brewing of beer; product specifications apply
Hops, extract solid	[*H. lupulus* (L.)]	2579	FDA §182.1	Natural flavor
Hops, oil	[*H. lupulus* (L.)]	2580	FDA §182.1	Natural flavor
Horehound, extract	Hoarhound [*Marrubium vulgare* Tourn. (L.)]	2581	FDA §182.1	GRAS, natural flavor and natural flavor extractive
Horsemint leaves, extract	[*Monarda punctata* (L.)]	2582	FDA §182.1	GRAS, natural flavor extractive
Horseradish	[*Armoracia lapthifolia* Gilib.]		FDA §182.1	GRAS, natural flavor

REGULATORY STATUS OF DIRECT FOOD ADDITIVES (continued)

Material	Alternate Name [Source]	FEMA No.	Regulation	Limitations
Hyacinth, absolute	[*Hyacinthus orientalis* (L.)]	—	FDA §172.510	Natural flavor
Hyacinth, flowers	[*Hyacinthus orientalis* (L.)]		FDA §172.510	Natural flavor
Hyacinthin	See phenylacetaldehyde			
Hydratropaldehyde	See 2-phenylpropionaldehyde			
Hydratropaldehyde dimethyl acetal	See 2-phenylpropionaldehyde dimethyl acetal			
Hydratropic alcohol	See β-methylphenethyl alcohol			
Hydratropic aldehyde dimethyl acetal	See 2-phenyl propionaldehyde dimethyl acetal			
Hydratropyl alcohol	See β-methylphenethyl alcohol			
Hydratropyl butyrate	See 2-phenylpropylbutyrate			
Hydratropyl isobutyrate	See 2-phenylpropyl isobutyrate			
Hydratropyl 2-methylpropanoate	See 2-phenylpropyl isobutyrate			
Hydrochloric acid	Muriatic acid		FDA §182.1	GRAS, buffer and neutralizing agent
			FDA §121.1031	Modifier for food starch
Hydrocinnamaldehyde	See 3-phenylpropionaldehyde			
Hydrocinnamic acid	See 3-phenylpropionic acid			
Hydrocinnamyl acetate	See 3-phenylpropyl acetate			
Hydrocinnamyl alcohol	See 3-phenyl-1-propanol			
Hydrocinnamyl caproate	See 3-phenylpropyl hexanoate			
Hydrocinnamyl cinnamate	See 3-phenylpropyl cinnamate			
Hydrocinnamyl formate	See 3-phenylpropyl formate			
Hydrocinnamyl hexanoate	See 3-phenylpropyl hexanoate			
Hydrocinnamyl isobutyrate	See 3-phenylpropyl isobutyrate			
Hydrocinnamyl isovalerate	See 3-phenylpropyl isovalerate			
Hydrocinnamyl methanoate	See 3-phenylpropyl formate			
Hydrocinnamyl 3-methylbutanoate	See 3-phenylpropyl isovalerate			
Hydrocinnamyl 2-methylpropanoate	See 3-phenylpropyl isobutyrate			

Hydrocinnamyl 3-phenylpropenoate	See 3-phenylpropyl cinnamate			
Hydrocinnamyl propionate	See 3-phenylpropyl propionate			
2-Hydrocinnamyl tetrahydrofuran	See 2-(3-phenylpropyl)-tetrahydrofuran			
Hydrocoumarin	See dihydrocoumarin			
Hydrocumin alcohol	See *p*-mentha-1,8-dien-7-ol			
Hydrogen peroxide	Hydrogen dioxide		FDA §182.1 FDA §121.1031 MID	GRAS, bleaching agent Source of active oxygen for modification of food starch Bleaching agent in tripe; must be removed from the product by rinsing with clear water
Hydrogen sulfide		—		
Hydrolyzed plant protein			MID	To flavor various meat products
Hydrolyzed protein			FDA §37.1	In canned tuna
α-Hydro-omega-hydroxy-poly (oxyethylene) poly (oxypropylene) Block copolymers poly (oxypropylene) 53—59 mol poly (oxyethylene) 14—16 mol MW range 3500—4125 Cloud point 9—12°C in 10% aqueous solution			121.1137	In combination with dioctyl sodium sulfosuccinate in fumaric acid acidulated dry beverage bases and fruit juice drinks when food standards do not apply. Finished beverages not to contain in excess of 10 ppm DSS-block polymer combination
α-Hydro-omega-hydroxy-poly (oxyethylene) poly (oxypropylene) Block copolymers poly (oxypropylene) 55—61 mol MW range 9760—13,200 Cloud point above 100°C in 1% aqueous solution			121.1235	As a solubilizing agent and stabilizer for flavor concentrates used in foods when standards do not apply. Weight must not exceed weight of flavoring oils in flavor concentrate

REGULATORY STATUS OF DIRECT FOOD ADDITIVES (continued)

Material	Alternate Name [Source]	FEMA No.	Regulation	Limitations
α-Hydro-omega-hydroxy-poly (oxyethylene) poly (oxypropylene) Block copolymers poly (oxypropylene) 51—57 mol Average MW 14,000 Cloud Point above 100°C in 1% aqueous solution			121.1235	For use in yeast-leavened bakery products for which food standards do not apply. Amount not to exceed 0.5% by weight of flour
Hydroquinone dimethyl ether	See *p*-dimethoxybenzene			
Hydroquinone monoethyl ether		—		
o-Hydroxyanisole	See guaiacol			
o-Hydroxybenzaldehyde	See salicylaldehyde			
2-Hydroxybenzaldehyde	See salicylaldehyde			
Hydroxybenzene	See phenol			
p-Hydroxybenzoic acid butyl ester	See butyl *p*-hydroxybenzoate			
p-Hydroxybenzyl acetone	See 4-(*p*-hydroxyphenyl)-2-butanone			
5-Hydroxy-2-benzyl-1,3-dioxane	See phenylacetaldehyde glyceryl acetal			
2-Hydroxybornane	See borneol			
2-Hydroxybutanedioic acid	See *l*-malic acid			
2-Hydroxy-3-butanethiol	See 2-mercapto-3-butanol			
3-Hydroxy-2-butanethiol	See 2-mercapto-3-butanol			
4-Hydroxybutanoic acid lactone	γ-Butyrolactone	3291		See *Food Technol.*, May 1972
1-Hydroxy-2-butanone		3173		

3-Hydroxy-2-butanone	See acetoin			
2-Hydroxycamphane	See borneol			
endo-2-Hydroxycamphane	See borneol			
Hydroxycitronellal	3,7-Dimethyl-7-hydroxyoctanal	2583	FDA §172.515	Synthetic flavor
Hydroxycitronellal diethyl acetal		2584	FDA §172.515	Synthetic flavor
Hydroxycitronellal dimethyl acetal		2585	FDA §172.515	Synthetic flavor
Hydroxycitronellal	3,7-Dimethyl-1,7-octanediol	2586	FDA §172.515	Synthetic flavor
2-Hydroxy-2-cyclohexen-1-one		3458		
2-Hydroxy-*p*-cymene	See carvacrol			
3-Hydroxy-*p*-cymene	See thymol			
4-Hydroxydecanoic acid, γ-lactone	See γ-decalactone			
5-Hydroxydecanoic acid, δ-lactone	See δ-decalactone			
o-Hydroxydihydrocinnamic acid lactone	See dihydrocoumarin			
Hydroxydihydrocitronellol	See hydroxycitronellol			
6-Hydroxydihydrotheaspirane	6-Hydroxy-2,6,10,10-tetramethyl-1-oxaspiro-[4.5]decane	3549		See *Food Technol.*, February 1978
2-Hydroxy-1,3-dimethoxybenzene	See 2,6-dimethyoxyphenol			
1-Hydroxy-2,5-dimethylbenzene	See 2,5-xylenol			
2-Hydroxy-1,3-dimethylbenzene	See 2,6-xylenol			
6-Hydroxy-3,7-dimethylcaprylic acid, ε-lactone	See 6-hydroxy-3,7-dimethyloctanoic acid lactone			
4-Hydroxy-2,5-dimethyl-3(2H)-furanone		3174		
6-Hydroxy-3,7-dimethyloctanoic acid lactone		3355		
4-Hydroxydodecanoic acid, γ-lactone	See γ-dodecalactone			
5-Hydroxydodecanoic acid, δ-lactone	See δ-dodecalactone			

REGULATORY STATUS OF DIRECT FOOD ADDITIVES (continued)

Material	Alternate Name [Source]	FEMA No.	Regulation	Limitations
4-Hydroxyethylbenzene	See *p*-ethylphenol			
4-(α-Hydroxyethyl)toluene	See *p*, α-dimethylbenzyl alcohol			
Hydroxyheptane	See heptyl alcohol			
2-Hydroxyheptane	See 2-heptanol			
4-Hydroxyheptanoic acid, γ-lactone	See γ-heptalactone			
16-Hydroxy-6-hexadecenoic acid, ω-lactone	See ω-6-hexadecenlactone			
4-Hydroxyhexanoic acid, γ-lactone	See γ-hexalactone			
5-Hydroxyhexanoic acid lactone	See δ-hexalactone			
1-Hydroxy-2-isopropylbenzene	See 2-isopropylphenol			
Hydroxylated lecithin			FDA §121.1027	Emulsifier in foods, except standardized foods that do not provide for such use; product specifications apply
			FDA §121.1099	In defoaming agents limited to use in processing beet sugar and yeast
4-Hydroxy-3-methoxy-1-allylbenzene	See eugenol			
4-Hydroxy-3-methoxybenzaldehyde	See vanillin			
1-Hydroxy-2-methoxybenzene	See guaiacol			
4-Hydroxy-3-methoxybenzylacetone	See zingerone			
4-Hydroxy-3-methoxybenzyl alcohol	See vanillyl alcohol			
N-(4-Hydroxy-3-methoxybenzyl)-8-methyl-6-nonenamide		3404		

N-(4-Hydroxy-3-methoxybenzyl)nonanamide	Pelargonyl vanillylamide; nonanoyl 4-hydroxy-3-methoxybenzylamide	2787	FDA §172.515	Synthetic flavor
1-Hydroxy-2-methoxy-4-ethylbenzene	See 4-ethylguaiacol			
1-Hydroxy-2-methoxy-4-methylbenzene	See 2-methoxy-4-methylphenol			
4-Hydroxy-3-methoxy-1-methylbenzene	See 2-methoxy-4-methylphenol			
4-(4-Hydroxy-3-methoxyphenyl)-2-butanone	See zingerone			
4-(4-Hydroxy-3-methoxyphenyl)but-3-en-2-one	See vanillylidene acetone			
2-(4-Hydroxy-3-methoxyphenyl)ethyl methyl ketone	See zingerone			
4-Hydroxy-3-methoxyphenylmethanol	See vanillyl alcohol			
1-Hydroxy-2-methoxy-4-propenylbenzene	See eugenol			
1-Hydroxy-2-methoxy-4-propen-1-ylbenzene	See iso-eugenol			
4-Hydroxy-3-methoxy-1-propen-1-ylbenzene	See iso-eugenol			
4-Hydroxy-3-methoxystyrene	See 2-methoxy-4-vinylphenol			
Hydroxymethyl anethole	See propenyl guaethol			
1-Hydroxy-2-methylbenzene	See *o*-cresol			
1-Hydroxy-3-methylbenzene	See *m*-cresol			
1-Hydroxy-4-methylbenzene	See *p*-cresol			
2-Hydroxy-1-methylbenzene	See *o*-cresol			
5-Hydroxymethyl-2-benzyl-1,3-dioxolane	See phenylacetaldehyde glyceryl acetal			
2-Hydroxymethyl-6,6-dimethylbicyclo(3.1.1)hept-2-enyl formate		3405		

REGULATORY STATUS OF DIRECT FOOD ADDITIVES (continued)

Material	Alternate Name [Source]	FEMA No.	Regulation	Limitations
2-Hydroxymethylfuran	See furfuryl alcohol			
2-Hydroxy-3-methyl-γ-2-hexenolactone	See 5-ethyl-3-hydroxy-4-methyl-2(5*H*)-furanone			
1-Hydroxymethyl-4-isopropenyl-1-cyclohexene	See *p*-mentha-1,8-dien-7-ol			
1-Hydroxy-2-methylmercaptobenzene	See *o*-(methylthio)-phenol			
3-(Hydroxymethyl)-2-octanone		3292		See *Food Technol.*, May 1972
4-Hydroxymethyl-2-phenyl-*m*-dioxolane	See benzaldehyde glyceryl acetal			
2-Hydroxy-2-methyl-1-phenylpropane	See α,α-dimethylphenethyl alcohol			
3-Hydroxy-2-methyl-4*H*-pyran-4-one	See maltol			
3-Hydroxy-2-methyl-γ-pyrone	See maltol			
Hydroxynonanoic acid, δ-lactone		3356		
4-Hydroxynonanoic acid, γ-lactone	See γ-nonalactone			
5-Hydroxynonanoic acid lactone	See hydroxynonanoic acid, δ-lactone			
1-Hydroxy-3-nonanone acetate	See 3-nonanon-1-yl acetate			
4-Hydroxyoctanoic acid, γ-lactone	See γ-octalactone			
5-Hydroxyoctanoic acid lactone	See δ-octalactone			
5-Hydroxy-4-octanone	Butyroin	2587	FDA §172.515	Synthetic flavor
γ-Hydroxy-β-oxobutane	See acetoin			
15-Hydroxypentadecanoic acid, ω-lactone	See ω-pentadecalactone			

4-Hydroxypentanoic acid, γ-lactone	See γ-valerolactone			
3-Hydroxy-2-pentanone	Acetyl ethyl carbinol	3550		See *Food Technol.*, February 1978
4-Hydroxy-3-pentenoic acid lactone	5-Methyl-2(3*H*)-furanone	3293		See *Food Technol.*, May 1972
2-Hydroxy-2-phenylacetophenone	See benzoin			
α-Hydroxy-α-phenylacetophenone	See benzoin			
1-*p*-Hydroxyphenyl-3-butanone	See 4-(*p*-hydroxyphenyl)-2-butanone			
4-(*p*-Hydroxyphenyl)-2-butanone	*p*-Hydroxybenzyl acetone	2588	FDA §172.515	Synthetic flavor
5-Hydroxy-2-phenyl-1,3-dioxane	See benzaldehyde glyceryl acetal			
1-(2-Hydroxyphenyl)-propane	See *o*-propylphenol			
2-Hydroxy-1,2,3-propanetricarboxylic acid	See citric acid			
2-Hydroxypropanoic acid	See lactic acid			
α-Hydroxypropanoic acid	See lactic acid			
2-Hydroxypropionic acid	See lactic acid			
α-Hydroxypropylbenzene	See 1-phenyl-1-propanol			
Hydroxypropylcellulose			FDA §121.1160	Emulsifier, film former, protective colloid, stabilizer, suspending agent, or thickener in foods; product specifications apply
Hydroxypropylmethylcellulose			FDA §121.1021	Emulsifier, film former, protective colloid, stabilizer, suspending agent, or thickener, except in confectionery and standardized foods that do not provide for use; product specifications apply
γ-Hydroxypropylmethyl sulfide	See 3-(methylthio)propanol			
1-Hydroxysuccinic acid	See *l*-malic acid			
5-Hydroxytetradecanoic acid lactone	See Δ-tetradecalactone			

REGULATORY STATUS OF DIRECT FOOD ADDITIVES (continued)

Material	Alternate Name [Source]	FEMA No.	Regulation	Limitations
6-Hydroxy-2,6,10,10-tetramethyl-1-oxaspiro-[4.5]decan	See 6-hydroxydihydrotheaspirane			
o-Hydroxytoluene	See *o*-cresol			
p-Hydroxytoluene	See *p*-cresol			
3-Hydroxytoluene	See *m*-cresol			
α-Hydroxytoluene	See benzyl alcohol			
β-Hydroxytricarboxylic acid	See citric acid			
4-Hydroxy-2,6,6-trimethylbicyclo[3.1.1]-hept-2-ene	See verbenol			
2-Hydroxy-3,5,5-trimethyl-2-cyclohexenone		3459		
4-Hydroxyundecanoic acid, γ-lactone	See γ-undecalactone			
4-Hydroxyundecanoic acid lactone	See γ-undecalactone			
5-Hydroxyundecanoic acid lactone	δ-Undecalactone	3294		See *Food Technol.*, May 1972
Hyponitrous acid anhydride	See nitrous oxide			
Hyssop	[*Hyssopus officinalis* (L.)]	2589	FDA §182.1	GRAS, natural flavor and natural flavor extractive
Hyssop, extract	[*H. officinalis* (L.)]	2590	FDA §182.1	Natural flavor
Hyssop, oil	[*H. officinalis* (L.)]	2591	FDA §182.1	Natural flavor
Iceland moss	[*Cetraria islandica* (L.) Ach.]	—	FDA §172.510	Natural flavor in alcoholic beverages only
Imidole	See pyrrole			
Immortelle, extract	Helichrysum [*Helichrysum angustifolium* D.C.]	2592	FDA §182.1	GRAS, natural flavor extractive
Imperatoria	[*Peucedanum ostruthium* (L.) Koch; *Imperatoria ostruthium* (L.)]	—	FDA§172.510	Natural flavor in alcoholic beverages only

Impruvol	See butylated hydroxytoluene			
Indian gum	See ghatti, gum [*Anogeissus latifolia* Wall.]			
Indole	2,3-Benzopyrrole	2593	FDA §172.515	Synthetic flavor
Inositol	*mesa*-Inositol		FDA §182.1	GRAS, nutrient and/or dietary supplement
Invert sugar	Nulomoline		FDA §8.303	In production of caramel
Ionol C.P.	See butylated hydroxytoluene			
α-Ionone	4-(2,6,6-Trimethyl-2-cyclohexen-1-yl)-3-buten-2-one	2594	FDA §172.515	Synthetic flavor
β-Ionone	4-(2,6,6-Trimethyl-1-)cyclohexen-1-yl)-3-buten-2-one	2595	FDA §172.515	Synthetic flavor
γ-Ionone	4-(2,2-Dimethyl-6-methylene-cyclohexyl)-3-buten-2-one	3175		
I.P.A.	See isopropyl alcohol			
β-Iraldeine	See methyl-β-ionone			
Iraldeine gamma	See α-iso-methylionone			
Irish moss, extract	Carrageenan	2596	FDA §182.1	
α-Irisone	See α-ionone			
β-Irisone	See β-ionone			
Iron ammonium citrate	Green ferric ammonium citrate		FDA §121.1190	Anticaking agent in salt; 25 ppm max in finished salt; product specifications apply
Iron-choline citrate complex			FDA §121.1100	Source of iron in foods for special dietary use, when made by reacting ca. equimolar quantities of ferric hydroxide, choline, and citric acid
cis-(2,6)-*cis*-(2-(1),2(2))-α-Irone	See α-irone			
α-Irone	4-(2,5,6,6-Tetramethyl-2-cyclohexen-1-yl)-3-buten-2-one; 6-methylionone	2597	FDA §172.515	Synthetic flavor
Iron, reduced			FDA §182.1	GRAS, nutrient and/or dietary supplement
Isoacetophorone	See isophorone			
Isoamyl acetate	Amylacetic ester	2055	FDA §172.515	Synthetic flavor

REGULATORY STATUS OF DIRECT FOOD ADDITIVES (continued)

Material	Alternate Name [Source]	FEMA No.	Regulation	Limitations
Isoamyl acetoacetate	—	—	FDA §172.515	Synthetic flavor
	iso-Amyl acetoacetate	3551		See *Food Technol.*, February 1978
	Isoamyl 3-oxobutanoate	3551		See *Food Technol.*, February 1978
	3-Methylbutyl 3-oxobutanoate	3551		See *Food Technol.*, February 1978
Isoamyl alcohol	Isopentyl alcohol; 3-methyl-1-butanol	2057	FDA §172.515	Synthetic flavor
Isoamyl aldehyde	See 3-methylbutyraldehyde			
Isoamylamine	See isopentylamine			
Isoamyl benzoate	Isopentyl benzoate	2058	FDA §172.515	Synthetic flavor
Isoamyl butanoate	See isoamyl butyrate			
Isoamyl butyrate	Isopentyl butyrate	2060	FDA §172.515	Synthetic flavor
Isoamyl caproate	See isoamyl hexanoate			
Isoamyl caprylate	See isoamyl octanoate			
Isoamyl cinnamate	Isopentyl cinnamate	2063	FDA §172.515	Synthetic flavor
Isoamyl dodecanoate	See isoamyl laurate			
Isoamyl ethanoate	See isoamyl acetate			
Isoamyl formate	Isopentyl formate	2069	FDA §172.515	Synthetic flavor
Isoamyl 2-furanbutyrate	α-Isoamyl furfurylpropionate	2070	FDA §172.515	Synthetic flavor
Isoamyl 2-furanpropionate	α-Isoamyl furfurylacetate	2071	FDA §172.515	Synthetic flavor
α-Isoamyl-furfurylacetate	See isoamyl 2-furanpropionate			
α-Isoamyl furfurylpropionate	See isoamyl 2-furanbutyrate			
Isoamyl hexanoate	Isopentyl hexanoate	2075	FDA §172.515	Synthetic flavor
Isoamyl *o*-hydroxybenzoate	See isoamyl salicylate			
Isoamyl isobutyrate		—	FDA §172.515	Synthetic flavor
Isoamyl isopentanoate	See isoamyl isovalerate			
Isoamyl isovalerate	Isoamyl isovalerianate	2085	FDA §172.515	Synthetic flavor
Isoamyl isovalerianate	See isoamyl isovalerate			
Isoamyl β-ketobutyrate	See isoamyl acetoacetate			
Isoamyl laurate	Isoamyl dodecanoate; isopentyl laurate	2077	FDA §172.515	Synthetic flavor
Isoamyl methanoate	See isoamyl formate			
Isoamyl 2-methylbutanoate	See isoamyl 2-methylbutyrate			

Isoamyl 3-methylbutanoate	See isoamyl isovalerate			
Isoamyl 2-methylbutyrate		3505	FDA §172.515	Synthetic flavor
Isoamyl 2-methylpropanoate	See 3-methylbutyl 2-methylpropanoate			
Isoamyl 2-methylpropanoate	See isoamyl isobutyrate			
Isoamyl nonanoate	Isopentyl nonanoate	2078	FDA §172.515	Synthetic flavor
Isoamyl nonylate	See isoamyl nonanoate			
Isoamyl octanoate	Isopentyl octanoate	2080	FDA §172.515	Synthetic flavor
Isoamyl octylate	See isoamyl octanoate			
Isoamyl 3-oxobutanoate	See isoamyl acetoacetate			
Isoamyl pelargonate	See isoamyl nonanoate			
Isoamyl pentanoate	See isoamyl valerate			
Isoamyl phenylacetate	Isopentyl phenylacetate	2081	FDA §172.515	Synthetic flavor
Isoamyl β-phenylacrylate	See isoamyl cinnamate			
Isoamyl propanoate	See isoamyl propionate			
Isoamyl propionate	Isopentyl propionate	2082	FDA §172.515	Synthetic flavor
Isoamyl pyruvate	Isopentyl pyruvate	2083	FDA §172.515	Synthetic flavor
Isoamyl salicylate	Isopentyl salicylate	2084	FDA §172.515	Synthetic flavor
Isoamyl α-toluate	See isoamyl phenylacetate			
Isoamyl valerate		—		
Isoanethole	See estragole			
Isoascorbic acid	See erythorbic acid			
Isoborneol	exo-2-Bornanol	2158	FDA §172.515	Synthetic flavor
Isobornyl acetate		2160	FDA §172.515	Synthetic flavor
Isobornyl alcohol	See isoborneol			
Isobornyl ethanoate	See isobornyl acetate			
Isobornyl formate		2162	FDA §172.515	Synthetic flavor
Isobornyl isopentanoate	See isobornyl isovalerate			
Isobornyl isovalerate		2166	FDA §172.515	Synthetic flavor
Isobornyl isovalerianate	See isobornyl isovalerate			
Isobornyl methanoate	See isobornyl formate			
Isobornyl 3-methylbutanoate	See isobornyl isovalerate			
Isobornyl propionate		2163	FDA §172.515	Synthetic flavor
Isobutanol	See isobutyl alcohol			
2-Isobutoxynaphthalene	See β-naphthyl isobutyl ether			
Isobutyl acetate		2175	FDA §172.515	Synthetic flavor

REGULATORY STATUS OF DIRECT FOOD ADDITIVES (continued)

Material	Alternate Name [Source]	FEMA No.	Regulation	Limitations
Isobutyl acetoacetate		2177	FDA §172.515	Synthetic flavor
Isobutyl alcohol	Isobutanol	2179	FDA §172.515	Synthetic flavor
			FDA §8.390 and §8.6200	Diluent in inks for marking gum, confectionery, and food supplements in tablet form and in inks for branding pharmaceutical forms; no residue in finished confectionery, gum, or tablets
Isobutyl aldehyde	See isobutyraldehyde			
Isobutyl angelate	Isobutyl *cis*-2-methyl-2-butenoate	2180	FDA §172.515	Synthetic flavor
Isobutyl anthranilate		2182	FDA §172.515	Synthetic flavor
Isobutyl benzoate		2185	FDA §172.515	Synthetic flavor
Isobutyl benzyl carbinol	See α-isobutylphenethyl alcohol			
Isobutyl benzyl ketone	See 4-methyl-1-phenyl-2-pentanone			
Isobutyl butanoate	See isobutyl butyrate			
Isobutyl 2-butenoate		3432		
Isobutyl butyrate		2187	FDA §172.515	Synthetic flavor
Isobutyl caproate	See isobutyl hexanoate			
Isobutyl carbinol	See isoamyl alcohol			
isobutyl carbylamine	See isopentylamine			
Isobutyl cinnamate		2193	FDA §172.515	Synthetic flavor
Isobutyl crotonate	See isobutyl 2-butenoate			
Isobutylene-isoprene copolymer	Butyl rubber		FDA §121.1059	Synthetic masticatory substance in chewing-gum base
Isobutylene resin	Polyisobutylene		FDA §121.1059	Synthetic masticatory substance in chewing-gum base; product specifications apply
Isobutyl ethanoate	See isobutyl acetate			
Isobutyl formate	Tetryl formate	2197	FDA §172.515	Synthetic flavor/adjuvant
Isobutyl formic acid	See isovaleric acid			
Isobutyl 2-furanpropionate	Isobutyl furylpropionate	2198	FDA §172.515	Synthetic flavor/adjuvant
Isobutyl furylpropionate	See isobutyl 2-furanpropionate			
Isobutyl heptanoate		2200	FDA §172.515	Synthetic flavor/adjuvant
Isobutyl heptoate	See isobutyl heptanoate			

Isobutyl hexanoate		2202	FDA §172.515	Synthetic flavor/adjuvant
Isobutyl *o*-hydroxybenzoate	See isobutyl salicylate			
Isobutyl isobutyrate		2189	FDA §172.515	Synthetic flavor
Isobutyl isovalerate	See 2-methylpropyl 3-methylbutyrate			
Isobutyl 3-ketobutanoate	See isobutyl acetoacetate			
Isobutyl 3-ketobutyrate	See isobutyl acetoacetate			
Isobutyl ketone	See 2,6-dimethyl-4-heptanone			
Isobutyl methanoate	See isobutyl formate			
2-Isobutyl-3-methoxypyrazine		3132		*10 ppm max in finished food*
Isobutyl 2-*N*-methylaminobenzoate	See isobutyl *N*-methylanthranilate			
Isobutyl *N*-methylanthranilate			Under FEMA review	
Isobutyl *cis*-2-methyl-2-butenoate	See isobutyl angelate			
Isobutyl *cis*-2-methylcyclohexanol			FDA §172.515	Synthetic flavor
2-Isobutyl-5-methylcyclohexanol	See *d*-Neomenthol			
Isobutyl methyl ketone	See 4-methyl-2-pentanone			
Isobutyl 2-methylpropanoate	See isobutyl isobutyrate			
2-Isobutyl-3-methylpyrazine	2-Methyl-3-isobutylpyrazine	3133		10 ppm max in finished food
Isobutyl *β*-naphthyl *ether*	See *β*-*naphthyl isobutyl ether*			
Isobutyl 3-oxobutanoate	See isobutyl acetoacetate			
α-Isobutylphenethyl alcohol	Isobutyl benzyl carbinol; 4-methyl-1-phenyl-2-pentanol	2208	FDA §172.515	Synthetic flavor
Isobutyl phenylacetate		2210	FDA §172.515	Synthetic flavor
Isobutyl *β*-phenylacrylate	See isobutyl cinnamate			
Isobutyl 3-phenylpropenoate	See isobutyl cinnamate			
Isobutyl propanoate	See isobutyl propionate			
Isobutyl propionate		2212	FDA §172.515	Synthetic flavor
2-Isobutyl pyridine	See 2-(2-methylpropyl)pyridine			
3-Isobutyl pyridine	See 3-(2-methylpropyl)pyridine			

REGULATORY STATUS OF DIRECT FOOD ADDITIVES (continued)

Material	Alternate Name [Source]	FEMA No.	Regulation	Limitations
Isobutyl salicylate		2213	FDA §172.515	Synthetic flavor
2-Isobutyl thiazole		3134		
Isobutyl α-toluate	See isobutyl phenylacetate			
Isobutyraldehyde	2-Methyl propanal	2220	FDA §172.515	Synthetic flavor
Isobutyric acid	Isopropylformic acid	2222	FDA §172.515	Synthetic flavor
Isobutyric acid, benzyl ester	See benzyl isobutyrate			
Isobutyric aldehyde	See isobutyraldehyde			
Isocaproic acid	See 4-methylpentanoic acid			
Isodecenoic acid vanillylamide	See *N*-(4-hydroxy-3-methoxybenzyl)-8-methyl-6-nonenamide			
Isodecylaldehyde	See 2,6-dimethyl octanal			
Isoeugenol	2-Methoxy-4-propenylphenol; iso-eugenol	2468	FDA §172.515	Synthetic flavor
Isoeugenol acetate	See iso-eugenyl acetate			
Isoeugenol methyl ether	See isoeugenyl methyl ether			
Isoeugenyl acetate	See iso-eugenyl acetate			
Isoeugenyl benzyl ether	Benzyl isougenol	—	FDA §172.515	Synthetic flavor/adjuvant
Isoeugenyl ethyl ether	2-Ethoxy-5-propenylanisole; ethyl isoeugenol	2472	FDA §172.515	Synthetic flavor
Isoeugenyl formate	2-Methoxy-4-propenylphenyl formate	2474	FDA §172.515	Synthetic flavor
Isoeugenyl methyl ether	Methyl isoeugenol; 4-propenyl veratrole	2476	FDA §172.515	Synthetic flavor
Isoeugenyl phenylacetate	2-Methoxy-4-propenylphenyl phenylacetate	2477	FDA §172.515	Synthetic flavor
Isoeugenyl α-toluate	See isoeugenyl phenylacetate			
Isohexanoic acid	See 4-methylpentanoic acid			
Isohexanone	See 4-methyl-2-pentanone			
Isojasmone	Mixture of 2-hexylidenecyclopentanone and 2-hexyl-2-cyclopenten-1-one	—	FDA §172.515	Synthetic flavor/adjuvant
	Dihydrojasmone	3552		See *Food Technol.*, February 1978
	2-Hexyl-2-cyclopenten-1-one and 2-hexylidene-cyclopentanone mixture	3552		See *Food Technol.*, February 1978

Isolated soy protein			MID	To bind and extend imitation sausage, nonspecific loaves, soups, and stews
			MID	Binder, alone or with other binders: 8% max in chili con carne 12% max in spaghetti with meatballs and sauce, spaghetti with meat and sauce, and similar products 2% max (3.5% of other binders) in sausage 2% max (3.5% of other binders) in cheesefurters and similar products No limit for pork with barbeque sauce or beef with barbeque sauce Isolated soy protein must contain 0.1% food-grade titanium dioxide as a tracer and must be declared on labels
Isoleucine (*l*- and *dl*-forms)			FDA §182.1	GRAS, nutrient and/or dietary supplement as the free acid, hydrochloride salt, hydrate, or anhydrous form
D,L-Isoleucine		3295	FDA §182.1	Natural flavor; see *Food Technol.*, May 1972
d,l-Isomerthone		3460		
α-Isomethylionone	Methyl γ-ionone; 4(2,6,6-trimethyl-2-cyclohexen-1-yl)-3-methyl-3-buten-2-one; α-iso-methylionone	2714	FDA §172.515	Synthetic flavor
Isopentaldehyde	See 3-methylbutyraldehyde			
Isopentyl acetoacetate	See isoamyl acetoacetate			
Isopentyl alcohol	See isoamyl alcohol			
Isopentylamine	Isoamylamine; 3-methylbutylamine	3219		
Isopentyl benzoate	See isoamyl benzoate			
Isopentyl butanoate	See isoamyl butyrate			
Isopentyl butyrate	See isoamyl butyrate			
Isopentyl caproate	See isoamyl hexanoate			
Isopentyl cinnamate	See isoamyl cinnamate			
Isopentyl dodecanoate	See isoamyl laurate			
Isopentyl dodecylate	See isoamyl laurate			
Isopentyl formate	See isoamyl formate			
Isopentyl 2-furanbutyrate	See isoamyl 2-furanbutyrate			

REGULATORY STATUS OF DIRECT FOOD ADDITIVES (continued)

Material	Alternate Name [Source]	FEMA No.	Regulation	Limitations
Isopentyl 2-furanpropionate	See isoamyl 2-furanpropionate			
Isopentyl hexanoate	See isoamyl hexanoate			
Isopentyl o-hydroxybenzoate	See isoamyl salicylate			
Isopentyl isobutyrate	See 3-methylbutyl 2-methylpropanoate			
Isopentyl isobutyrate	See isoamyl isobutyrate			
Isopentyl isopentanoate	See isoamyl isovalerate			
Isopentyl isovalerate	See isoamyl isovalerate			
Isopentyl β-ketobutyrate	See isoamyl acetoacetate			
Isopentyl laurate	See isoamyl laurate			
Isopentyl methanoate	See isoamyl formate			
Isopentyl 2-methylbutanoate	See 3-methylbutyl 2-methylbutanoate			
Isopentyl 3-methylbutanoate	See isoamyl isovalerate			
Isopentyl 2-methylbutyrate	See isoamyl 2-methylbutyrate			
Isopentyl 2-methylpropanoate	See 3-methylbutyl 2-methylpropanoate			
Isopentyl 2-methylpropanoate	See isoamyl isobutyrate			
Isopentyl nonanoate	See isoamyl nonanoate			
Isopentyl nonylate	See isoamyl nonanoate			
Isopentyl octanoate	See isoamyl octanoate			
Isopentyl octylate	See isoamyl octanoate			
Isoentyl 3-oxobutanoate	See isoamyl acetoacetate			
Isopentyl pelargonate	See isomayl nonanoate			
Isopentyl phenylacetate	See isoamyl phenylacetate			
Isopentyl β-phenylacrylate	See isoamyl cinnamate			
Isopentyl 3-phenylpropenoate	See isoamyl cinnamate			
Isopentyl propanoate	See isoamyl propionate			

Isopentyl propionate	See isoamyl propionate			
Isopentyl pyruvate	See isoamyl pyruvate			
Isopentyl salicylate	See isoamyl salicylate			
Isophorone	3,5,5-Trimethyl-2-cyclohexen-1-one	3553		See *Food Technol.*, February 1978
Isopropanol	See isopropyl alcohol			
4-Isopropenyl-1-cyclohexenecarbinol	See *p*-mentha-1,8-dien-7-ol			
4-Isopropenyl-1-cyclohexene carbinol	See perillyl acetate			
4-Isopropenyl-1-cyclohexene carbinyl acetate	See *p*-mentha-1,8-dien-7-yl acetate			
4-Isopropenyl-1-cyclohexene-1-carboxaldehyde	See perillaldehyde and *p*-mentha-1,8-dien-7-al			
2-Isopropenyl-1,4-diazine	See isopropenylpyrazine			
4-Isopropenyl-1-methyl-1-cyclohexanol	See *p*-menth-8-en-1-ol			
3-Isopropenyl-6-methylcyclohexanone	See *p*-menth-8-en-2-one			
Isopropenylpyrazine	2-Isopropenyl-1,4-diazine; 2(1-methylvinyl) pyrazine	3296		See *Food Technol.*, May 1972
p-Isopropenyltoluene	See *p*-α-dimethylstyrene			
Isopropyl acetate		2926	FDA §172.515	Synthetic flavor
Isopropylacetic acid	See isovaleric acid			
Isopropylacetone	See 4-methyl-2-pentanone			
p-Isopropylacetophenone		2927	FDA §172.515	Synthetic flavor
p-Isopropylacetylbenzene	See *p*-isopropylacetophenone			
Isopropyl alcohol	Isopropanol	2929	FDA §172.515	Synthetic flavor/adjuvant
			FDA §8.390 and §8.6200	Diluent in inks for marking gum and confectionery, inks for branding pharmaceutical forms, and ingested drugs
			FDA §8.305	Solvent for annatto extract; residues, 50 ppm max
p-Isopropylbenzaldehyde	See cuminaldehyde			
4-Isopropylbenzenecarboxaldehyde	See cuminaldehyde			
Isopropyl benzoate		2932	FDA §172.515	Synthetic flavor
p-Isopropylbenzyl alcohol	Cuminic alcohol; *p*-cymen-7-ol	2933	FDA §172.515	Synthetic flavor
Isopropyl butanoate	See isopropyl butyrate			

REGULATORY STATUS OF DIRECT FOOD ADDITIVES (continued)

Material	Alternate Name [Source]	FEMA No.	Regulation	Limitations
Isopropyl butyrate		2935	FDA §172.515	Synthetic flavor
Isopropyl caproate	See isopropyl hexanoate			
Isopropyl capronate	See isopropyl hexanoate			
Isopropyl carbinol	See isobutyl alcohol			
Isopropyl cinnamate		2939	FDA §172.515	Synthetic flavor
Isopropyl citrate			FDA §182.1	GRAS, sequestrant up to 0.02%
			MID	0.02% in oleomargarine to protect flavor
Isopropyl-*o*-cresol	See carvacrol			
6-Isopropyl-*m*-cresol	See thymol			
4-Isopropyl-1-cyclohexene-1-carboxaldehyde	See perillaldehyde			
Isopropyl formate		2944	FDA §172.515	Synthetic flavor
Isopropyl formic acid	See isobutyric acid			
Isopropyl hexanoate		2950	FDA §172.515	Synthetic flavor
Isopropyl hexylate	See isopropyl hexanoate			
p-Isopropylhydrocinnamaldehyde	See 3-(*p*-isopropyl)phenyl propionaldehyde			
Isopropylidene acetone	See 4-methyl-3-penten-2-one			
1-Isopropylidene-4-methyl-2-cyclohexanone	See pulegone			
Isopropyl isobutyrate		2937	FDA §172.515	Synthetic flavor
Isopropyl isopentanoate	See isopropyl isovalerate			
Isopropyl isovalerate		2961	FDA §172.515	Synthetic flavor
Isopropyl isovalerianate	See isopropyl isovalerate			
Isopropyl methanoate	See isopropyl formate			
2-Isopropyl-(3,5,or 6)-methoxypyrazine	See 2-methoxy-3(5 and 6)-isopropylpyrazine			
Isopropyl 3-methylbutanoate	See isopropyl isovalerate			
Isopropyl α-methylcrotonate	See isopropyl tiglate			

Isopropyl α-methylcrotonic acid	See isopropyl tiglate			
1-Isopropyl-4-methyl-2,4-cyclohexadiene	See α-phellandrene			
4-Isopropyl-1-methyl-1,5-cyclohexadiene	See α-phellandre			
5-Isopropyl-2-methyl-1,3-cyclohexadiene	See α-phellandre			
1-Isopropyl-4-methylcyclohexan-2-ol	See menthol			
1-Isopropyl-4-methyl-2-cyclohexanone	See isopulegone			
2-Isopropyl-5-methylcyclohexanone	See menthone			
4-Isopropyl-1-methylcyclohexan-3-one	See menthone			
3-Isopropyl-6-methylcyclohexanol	See *p*-menthan-2-ol			
4-Isopropyl-1-methyl-3-cyclohexen-1-ol				
4-Isopropyl-1-methyl-1-cyclohexen-3-one	See *d*-piperitone			
1-Isopropyl-4-methylcyclohex-2-yl acetate	See menthyl acetate			
1-Isopropyl-4-methylcyclohex-2-yl 3-methylbutanoate	See menthyl isovlerate			
2-Isopropyl-5-methyl-2-hexenal		3406		
Isopropyl methyl ketone	See 4-methyl-2,3-pentanedione			
2-Isopropyl-5-methylphenol	See thymol			
Isopropyl 2-methylpropanoate	See isopropyl isobutyrate			
2-Isopropyl-5-methylpyrazine	See 2-Methyl-5-isopropylpyrazine	3554		See *Food Technol.*, February 1978
Isopropyl myristate	Isopropyl tetradecanoate	3556		See *Food Technol.*, February 1978
o-Isopropylphenol	See 2-isopropylphenol			

REGULATORY STATUS OF DIRECT FOOD ADDITIVES (continued)

Material	Alternate Name [Source]	FEMA No.	Regulation	Limitations
2-Isopropylphenol		3461		
α-Isopropyl phenylacetaldehyde	See 3-methyl-2-phenylbutyraldehyde			
p-Isopropylphenylacetaldehyde	*p*-Cymen-7-carboxaldehyde	2954	FDA §172.515	Synthetic flavor
Isopropyl phenylacetate		2956	FDA §172.515	Synthetic flavor
Isopropyl 3-phenylpropenoate	See isopropylcinnamate			
3-(*p*-Isopropylphenyl) propionaldehyde	*p*-Isopropylhydrocinnamaldehyde; cuminyl acetaldehyde	2957	FDA §172.515	Synthetic flavor
Isopropyl propionate		2959	FDA §172.515	Synthetic flavor
Isopropyl tetradecanoate	See isoproyl myristate			
Isopropyl tiglate	Isopropyl-*α*-methyl crotonic acid	3229		
Isopropyl *α*-toluate	See isopropyl phenylacetate			
p-Isopropyltoluene	See *p*-cymene			
2-Isopropy-5-methylcyclohexanol	See *d*-neo-menthol			
Isopulegol	*p*-Menth-8-en-3-ol	2962	FDA §172.515	Synthetic flavor
Isopulegol acetate	See isopulegyl acetate			
Isopulegone	*p*-Menth-8-en-3-one	2964	FDA §172.515	Synthetic flavor
Isopulegyl acetate		2965	FDA §172.515	Synthetic flavor
Isoquinoline	2-Benzazine	2978	FDA §172.515	Synthetic flavor
Isothymol	See carvacrol			
Isovaleral	See 3-methylbutyraldehyde			
Isovaleraldehyde	See 3-methylbutyraldehyde			
Isovaleric acid	*β*-Methylbutyric acid	3102	FDA §172.515	Synthetic flavor
Isovaleric aldehyde	See 3-methylbutyraldehyde			
Isovalerone	See 2,6-dimethyl-4-heptanone			
Isovitamin C	See eythrobic acid			
Iva	Musk yarrow [*Achillea moschata* Jacq.]		FDA §172.510	Natural flavor in alcoholic beverages only
Iva, extract	[*A. moschata* Jacq.]			

Jaguar	See guar, gum [*Cyamopsis tetragnolobus*(L.)]			
Japan wax	Vegetable wax		FDA §8.390	Diluent in mixtures for coloring egg shells; no penetration of mixture through shell into egg
Jasmine	[*Jasminum officinale*(L.) and other *Jasminum* spp.]		FDA §182.1	GRAS, natural flavor extractive
Jasmine, absolute	[*J. grandiflorum*(L.)]	2598		
Jasmine aldehyde	See α-amylcinnamaldehyde			
Jasmine, concrete	[*J. grandiflorum*(L.)]	2599	FDA §182.1	Natural flavor
Jasmine, oil	[*J. grandiflorum*(L.)]	2600	FDA §182.1	Natural flavor
Jasmine, spiritus	[*J. grandiflorum*(L.)]	2601	FDA §182.1	Natural flavor
Jasmonal	See α-amylcinnamaldehyde			
Jasmonal H.	See α-hexylcinnamaldehyde			
Jasmone	See 3-Methyl-2-(2-pentenyl)-2-cyclopenten-1-one			
cis-Jasmone	3-Methyl-2-(2-pentenyl)-2-cyclopenten-1-one	3196	FDA §172.515	Synthetic flavor
iso-Jasmone	See isojasmone			
Jasmonyl	See 1,3-nonanediol acetate (mixed esters)			
Jelutong	[*Dyera costulata* Hook. f.; *D. lowii* Hook. f.]		FDA §121.1059	Natural (coagulated or concentrated latex) masticatory substance in chewing-gum base
Juniper, berries	[*Juniperus communis*(L.)]	2602	FDA §182.1	GRAS, natural flavor extractive
Juniper, extract	[*J. communis*(L.)]	2603		
Juniper, oil	[*J. communis*(L.)]	2604		
Juniper, tar	See cade oil			
Kadaya	See karaya, gum [*Sterculia urens* Roxb.]			
Karaya, gum	Sterculia gum [*S. urens* Roxb.]	2605	FDA §182.1	GRAS, stabilizer
Karion	See *d*-sorbitol			
Katilo	See karaya, gum [*S. urens* Roxb.]			
Kautschin	See *d*-limonene			
Kelp	Brown algae [Atlantic — *Laminaria digitatia; L. saccharina;* Pacific — *Macrocystis pyrifera* (L.) C. Agardh.]	2606	FDA §182.1	GRAS, natural substance and solvent-free natural extractive used in conjunction with natural flavorings

REGULATORY STATUS OF DIRECT FOOD ADDITIVES (continued)

Material	Alternate Name [Source]	FEMA No.	Regulation	Limitations
			FDA §121.1149	Source of iodine in foods for special dietary use; daily intake of iodine not to exceed 0.15 mg
Kentonarome	See methylcyclopentenolone			
3-Ketobutane-1-carboxylic acid	See levulinic acid			
2-Keto-4-butanethiol		3357		
3-Ketobutyraldehyde dimethyl acetal	See 3-oxobutanal, dimethyl acetal			
Ketone alcohol	See 3-octanon-1-ol			
Ketone C-7	See 2-heptanone			
β-Ketopropane	See acetone			
2-Ketopropionaldehyde	See pyruvaldehyde			
2-Ketopropionic acid	See pyruvic acid			
α-Ketopropionic acid	See pyruvic acid			
α-Ketopropionic aldehyde	See pyruvaldehyde			
2-Keto-1,7,7-trimethylnorcamphane	See *d*-camphor			
γ-Ketovaleric acid	See levulinic acid			
Kola nut, extract	Cola nut [*Cola acuminata* Schott and Endl. and other *Cola* spp.]	2607	FDA §182.1	GRAS, natural flavor extractive
Krameria, extract	See rhatany, extract [*Krameria* spp.]			
Kristallose	See saccharin, sodium salt			
Kullo	See karaya, gum [*Sterculia urens* Roxb.]			
Kuteeral	See karaya, gum [*S. urens* Roxb.]			
Labdanol	See isobutyl cinnamate			
Labdanum, absolute	[*Cistus ladaniferus* (L.); *C. creticus;* and other *Cistus* spp.]	2608	FDA §172.510	Natural flavor
Labdanum, oil	[*Cistus* spp.]	2609	FDA §172.510	
Labdanum, oleoresin	[*Cistus* spp.]	2610	FDA §172.510	
Lactic acid		2611	FDA §182.1	GRAS, miscellaneous and/or general-purpose food additive

Lactic acid, butyl ester, butyrate	See butyl butyryllactate		
Lactic acid-producing bacteria		FDA §17	In bakery products
		FDA §19	In specified cheeses
		MID	Harmless bacterial starters of the lactic-acid type, to develop flavor in dry sausage, pork roll, thuringer, Lebanon bologna, cervelat, and bologna, at 0.5% level
Lactose	4-*O*-β-D-Galactopyranosyl-D-glucose	FDA §8.303	In production of caramel
Lactylic esters of fatty acids		FDA §121.1048	Prepared from lactic acid and fatty acids meeting requirements of §121.1070(b); emulsifiers, plasticizers, or surface-active agents in bakery mixes, baked products, cake icings, fillings, and toppings, dehydrated fruits and vegetables, dehydrated fruit and vegetable juices, frozen desserts, liquid shortening for household use, pancake mixes, precooked instant rice, pudding mixes, solid-state, edible vegetable-fat-water emulsions as substitutes for milk or cream in beverage coffee, and with shortening and edible fats and oils when such are required in the foods listed above
Laevulic acid	See levulinic acid		
Laevulinic acid	See levulinic acid		
Lakes (FD&C)		FDA §8.501	Food, drug, and cosmetic use, conforming with specifications in §9.100
Lakes of color additives for food use		FDA §8.299	Product specifications apply
Larch gum	See arabinogalactan		
Lard		FDA-GRAS	
Larixinic acid	See maltol		
Laughing gas	See nitrous oxide		
Lauraldehyde	See lauric aldehyde		

REGULATORY STATUS OF DIRECT FOOD ADDITIVES (continued)

Material	Alternate Name [Source]	FEMA No.	Regulation	Limitations
Laurel	See bay, sweet			
Laurel berries	[*Laurus nobilis* (L.)]	2612		
Laurel camphor	See *d*-camphor			
Laurel leaves, extract	[*L. nobilis* (L.)]	2613		
Lauric acid	Dodecanoic acid	2614	FDA §121.1070	
Lauric aldehyde	Aldehyde C-12; dodecanal	2615	FDA §172.515	Synthetic flavor
Laurine	See hydroxycitronellal			
Laurostearic acid	See lauric acid			
Lauryl acetate	Dodecyl acetate	2616	FDA §172.515	Synthetic flavor
Lauryl alcohol	Alcohol C-12; dodecanol	2617	FDA §172.515	Synthetic flavor
Lauryl aldehyde	See lauric aldehyde			
Lauryl ethanoate	See lauryl acetate			
Lauryl isobutyrate	See dodecyl isobutyrate			
Lauryl 2-methylpropanoate	See dodecyl isobutyrate			
Lavandin	[*Lavandula hybrida* Rev., hybrid between *L. officinalis* Chaix and *L. latifolia* Vill.]	2618	FDA §182.1	GRAS, natural flavor extractive
Lavender	[*L. officinalis* Chaix]	2619	FDA §182.1	GRAS, natural flavor and natural flavor extractive
Lavender, absolute	[*L. officinalis* Chaix]	2620		
Lavender, concrete	[*L. officinalis* Chaix]	2621	FDA §182.1	Natural flavor
Lavender, oil	[*L. officinalis* Chaix]	2622	FDA §182.1	Natural flavor
Lavender, spike, oil	[*L. latifolia* Vill. (*L. spica* D.C.)]	3033	FDA §182.1	GRAS, natural flavor extractive
Leaf alcohol	See 3-hexen-1-ol			
Leaf aldehyde	See 2-hexenal			
Leche caspi	Sorva [*Couma macrocarpa* Barb. Rodr.]		FDA §121.1059	Natural (coagulated or concentrated latex) masticatory substance in chewing-gum base
Leche de vaca	[*Brosimum utile* (H.B.K.); *Pittier* and *Poulsenia* spp.; *Lacmellea standleyi* (Woodson) Monachino]		FDA §121.1059	Natural (coagulated or concentrated latex) masticatory substance in chewing-gum base
Lecithin	Phosphatidylcholine		FDA §182.1	GRAS, miscellaneous and/or general-

				purpose food additive
			MID	Emulsifier and antioxidant in oleomargarine and shortening
Lecithin, hydroxylated	See hydroxylated lecithin			
Leek oil		—		
Lemon balm	See balm			
Lemon, extract	[*Citrus limonum* (L.) Burm. f.]	2623	FDA §182.1	GRAS, natural flavor extractive
Lemon grass, oil	[*Cymbopogon citratus* D.C.; *C. flexuosus* (Nees.) Stapf.]	2624	FDA §182.1	GRAS, natural flavor extractive
Lemon, oil	[*Citrus limon* (L.) Burm. f.]	2625	FDA §182.1	Natural flavor
Lemon, oil, terpeneless	[*C. limon* (L.) Burm. f.]	2626	FDA §182.1	Natural flavor
Lemon peel	[*C. limonum* (L.) Burm. f.]	2625	FDA §182.1	GRAS, natural flavor extractive
Lemon verbena	[*Lippia citriodora* H.B.K.]	—	FDA §172.510	Natural flavor in alcoholic beverages only
Lepidine	6-Methylquinoline	2744	FDA §172.515	Synthetic flavor/adjuvant
Leucine (L- and DL-forms)			FDA §182.1	GRAS, nutrient and/or dietary supplement as the free acid, hydrochloride salt, hydrate, or anhydrous form
L-Leucine		3297	FDA §182.1	Natural flavor; see *Food Technol.*, May 1972
Leucoline	See quinoline			
Levulinic acid	3-Acetylpropionic acid	2627	FDA §172.515	Synthetic flavor
Licareol	See linalool			
Licorice, extract	Glycyrrhiza [*Glycyrrhiza glabra* (L.) and other *Glycyrrhiza* spp.]	2628	FDA §182.1	GRAS, natural flavor and natural flavor extractive
Licorice, extract, powder	[*G. glabra* (L.)]	2629	FDA §182.1	Natural flavor
Licorice, root	[*G. glabra* (L.)]	2630	FDA §182.1	Natural flavor
Ligusticum lactone	See 3-butylidenephthalide			
Limed rosin			FDA §8.390	Diluent in mixtures for coloring eggshells; no penetration of mixture into the egg
Lime, oil	[*Citrus aurantifolia* Swingle]	2631	FDA §182.1	GRAS, natural flavor extractive
Lime, oil, terpeneless	[*C. aurantifolia* (Christman) Swingle]	2632	FDA §182.1	Natural flavor
Limonene (*d*-, *l*-, and *dl*-)	*p*-Mentha-1,8-diene		FDA §182.1	GRAS, synthetic flavor /adjuvant
d-Limonene		2633	FDA §182.1	Natural flavor
Linaloe wood	[*Bursera delpechiana* Poiss. and other *Bursera* spp.]	2634	FDA §172.510	Natural flavor
Linalol	See linalool			

REGULATORY STATUS OF DIRECT FOOD ADDITIVES (continued)

Material	Alternate Name [Source]	FEMA No.	Regulation	Limitations
Linalool	3,7-Dimethyl-1,6-octadien-3-ol; linalol	2635	FDA §182.1	GRAS, synthetic flavor/adjuvant
Linalool acetate	See linalyl acetate			
Linalool benzoate	See linalyl benzoate			
Linalool butanoate	See linalyl butyrate			
Linalool formate	See linalyl formate			
Linalool isobutyrate	See linalyl isobutyrate			
Linalool 2-methylpropanoate	See linalyl isobutyrate			
Linalool octanoate	See linalyl octanoate			
Linalool oxide	*cis-* and *trans-*2-Vinyl-2-methyl-5-(1′-hydroxyl-1′-methylethyl)-tetrahydrofuran	—	FDA §172.515	Synthetic flavor
Linalool propanoate	See linalyl propionate			
Linalyl acetate	Bergamol	2636	FDA §182.1	GRAS, synthetic flavor/adjuvant
Linalyl anthranilate	3,7-Dimethyl-1,6-octadien-3-yl-anthranilate	2637	FDA §172.515	Synthetic flavor
Linalyl benzoate		2638	FDA §172.515	Synthetic flavor
Linalyl butyrate		2639	FDA §172.515	Synthetic flavor
Linalyl caproate	See linalyl hexanoate			
Linalyl capronate	See linalyl hexnoate			
Linalyl caprylate	See linalyl octanoate			
Linalyl cinnamate		2641	FDA §172.515	Synthetic flavor/adjuvant
Linalyl formate		2642	FDA §172.515	Synthetic flavor
Linalyl hexanoate		2643	FDA §172.515	Synthetic flavor
Linalyl hexoate	See linalyl hexanoate			
Linalyl isobutyrate		2640	FDA §172.515	Synthetic flavor
Linalyl isopentanoate	See linalyl isovalerate			
Linalyl isovalerate		2646	FDA §172.515	Synthetic flavor
Linalyl isovalerianate	See linalyl isovalerate			
Linalyl 3-methylbutanoate	See linalyl isovalerate			
Linalyl octanoate		2644	FDA §172.515	Synthetic flavor

Linalyl octoate	See linalyl octanoate			
Linalyl phenylacetate		3501		
Linalyl 3-phenylpropenoate	See linalyl cinnamate			
Linalyl propionate		2645	FDA §172.515	Synthetic flavor
Linalyl α-toluate	See linalyl phenylacetate			
Linden, flower	[*Tilia glabra* Vent.]	2647	FDA §182.1	GRAS, natural flavor and natural flavor extractive
Linden, leaf	[*Tilia* spp.]		FDA §172.510	Natural flavor in alcoholic beverages only
Linoleic acid	*cis*-9, *cis*-12-Octadecadienoic acid		FDA §182.1	GRAS, *nutrient and/or dietary supplement*, when prepared from edible fats and oils and free from chick-edema factor
Linoleic and linolenic acids	See 9,12-octadecadienoic acid (48%) and 9,12,15-octadecatrienoic acid (52%)			
Locust bean	[*Ceratonia siliqua* (L.)]		FDA §182.1	GRAS, natural flavor extractive
Locust bean gum	See carob bean, gum [*C. siliqua* (L.)]			
Lovage	[*Levisticum officinale* Koch.]	2649	FDA §172.510	Natural flavor
Lovage, extract	[*L. officinale* Koch.]	2650	FDA §172.510	
Lovage, oil	[*L. officinale* Koch.]	2651	FDA §172.510	
Lungmoss	Lungwort [*Sticta pulmonacea* Ach.]		FDA §172.510	Natural flavor
Lungwort	See lungmoss [*S. pulmonacea* Ach.]			
Lupulin	[*Humulus lupulus* (L.)]		FDA §182.1	GRAS, natural flavor extractive
2,6-Lutidine	See 2,6-Dimethylpyridine			
Lysine (L- and DL-forms)			FDA §182.1	GRAS, nutrient and/or dietary supplement as free acid, hydrochloride salt, hydrate, and anhydrous form
Mace (see also nutmeg)	[*Myristica fragrans* Houtt.]	2652	FDA §182.1	GRAS, natural flavor and natural flavor extractive
Mace, oil	[*M. fragrans* Houtt.]	2653	FDA §182.1	Natural flavor
Mace, oleoresin	[*M. fragrans* Houtt.]	2654	FDA §182.1	Natural flavor
Magnesium carbonate			FDA §182.1	GRAS, miscellaneous and/or general-purpose food additive
Magnesium hydroxide			FDA §182.1	GRAS, miscellaneous and/or general-purpose food additive
Magnesium oxide	Magnesia		FDA §182.1	GRAS, nutrient and/or dietary supplement

REGULATORY STATUS OF DIRECT FOOD ADDITIVES (continued)

Material	Alternate Name [Source]	FEMA No.	Regulation	Limitations
			FDA §182.1	GRAS, miscellaneous and/or general-purpose food additive
Magnesium phosphate (di- and tribasic)			FDA §182.1	GRAS, nutrient and/or dietary supplement
Magnesium salts of fatty acids			FDA §121.1071	Alone or with other salts of fatty acids; binder, emulsifier, and anticaking agent in food; product specifications apply
Magnesium silicate	Talc		FDA §22.8 and §22.9	In vanilla powder and in vanilla-vanillin powder
			FDA §182.1	GRAS, anticaking agent in table salt, up to 2%
Magnesium sulfate			FDA §182.1	GRAS, nutrient and/or dietary supplement
Maidenhair fern	[*Adiantum capillus-veneris* (L.)]	—	FDA §172.510	Natural flavor in alcoholic beverages only
Malic acid	Hydroxysuccinic acid		FDA §182.1	GRAS, miscellaneous and/or general-purpose food additive
l-Malic acid	Apple acid; *l*-hydroxysuccinic acid	2655	FDA §182.1	GRAS, natural flavor extractive
Malonic ester	See diethyl malonate			
Maltol	3-Hydroxy-2-methyl-4*H*-pyran-4-one	2656	FDA §172.515	Synthetic flavor
Malt syrup			FDA§8.303	In production of caramel
			MID	To flavor cured meat products at 2.5%
Maltyl isobutyrate		3462		
Mandarin, oil	Tangerine and petit-grain mandarin or tangerine [*Citrus reticulata* Blanco var. *mandarin*]	2657	FDA §182.1	GRAS, natural flavor extractive
Manganese chloride	Manganous fluoride		FDA §182.1	GRAS, nutrient and/or dietary supplement
Manganese citrate			FDA §182.1	GRAS, nutrient and/or dietary supplement
Manganese gluconate			FDA §182.1	GRAS, nutrient and/or dietary supplement

Manganese glycerophosphate			FDA §182.1	GRAS, nutrient and/or dietary supplement
Manganese hypophosphite			FDA §182.1	GRAS, nutrient and/or dietary supplement
Manganese oxide	Manganomanganic oxide		FDA §182.1	GRAS, nutrient and/or dietary supplement
Manganese sulfate			FDA §182.1	GRAS, nutrient and/or dietary supplement
Mannitol	D-Mannitol		FDA §182.1 FDA §121.1115	GRAS, nutrient and/or dietary supplement in special dietary foods, up to 5% In foods
Maple, mountain, extract	[*Acer spicatum* Lam.]	2757	FDA §172.510	Natural flavor
Marigold	See tagetes, oil [*Tagetes patula* (L.); *T. erecta* (L.); or *T. minuta* (L.) (*T. gladulifera* Schrank)]			
Marigold, pot	Calendula [*Calendula officinalis* (L.)]	2658	FDA §182.1	GRAS, natural flavor
Marjoram, oleoresin	[*Marjorana hortensis* Moench. ; *Origanum majorana* (L.); *O. vulgare* (L.)]			
Marjoram, pot	[*M. onites* (L.) Benth.]	2660	FDA §182.1	GRAS, natural flavor
Marjoram, seed	[*M. hortensis* Moench. *O. majorana* (L.)]	2661		
Marjoram, sweet	[*O. marjorana (L.); M. hortensis* Moench.]	2662	FDA §182.1	GRAS, natural flavor and natural flavor extractive
Marjoram, sweet, oil	[*M. hortensis* Moench.; *O. majorana* (L.)]	2663	FDA §182.1	Natural flavor
Marshmallow root	See althea root [*Althea officinalis* (L.)]			
Massaranduba balata	[*Manilkara huberi* (Ducke) Chevalier]		FDA §121.1059	Massaranduba balata and its solvent-free resin extract, natural (coagulated or concentrated latex) masticatory substance in chewing-gum base
Massaranduba chocolate	[*M. solimoesensis* Gilly]		FDA §121.1059	Natural (coagulated or concentrated latex) masticatory substance in chewing-gum base
Mastic gum		—		
Maté	[*Ilex paraguensis* St. Hil.]		FDA §182.1	GRAS, natural flavor extractive
Matsuka alcohol	See 1-octen-3-ol			

REGULATORY STATUS OF DIRECT FOOD ADDITIVES (continued)

Material	Alternate Name [Source]	FEMA No.	Regulation	Limitations
MEK	See 2-butanone			
Melaleucol	See nerolidol			
Melilotic acid lactone	See dihydrocoumarin			
Melilotus	Yellow melilot [*Melilotus officinalis* (L.) Lam.]		FDA §172.510	Natural flavor in alcoholic beverages only
Melissa, extract	See balm leaves, extract [*Melissa officinalis* (L.)]			
Melissa, oil	See balm, oil [*M. officinalis* (L.)]			
Melonal	See 2,6-dimethyl-5-heptenal			
Menthacamphor	See menthol			
p-Mentha-1,8-dien-7-al	Perillaldehyde	3557		See *Food Technol.*, February 1978
	4-Isopropenyl-1-cyclohexene-1-carboxaldehyde	3557		See *Food Technol.*, February 1978
p-Mentha-1,3-diene	Tirpilene	3558		See *Food Technol.*, February 1978
	α-Terpinene	3558		See *Food Technol.*, February 1978
	1-Methyl-4-isopropyl-1,3-cyclohexadiene	3558		See *Food Technol.*, February 1978
p-Mentha-1,4-diene	1-Methyl-4-isopropyl-1,4-cyclohexadiene	3559		See *Food Technol.*, February 1978
	γ-Terpinene	3559		See *Food Technol.*, February 1978
Menthadien-7-carbinol	See *p*-mentha-1,8-dien-7-ol			
Menthadien-7-carbinyl acetate	See perillyl acetate			
p-1,8(9)-Menthadiene	See *d*-limonene			
d-*p*-Mentha-1,8-diene	See *d*-limonene			
p-Mentha-1, 3-diene	See α-terpinene			
p-Mentha-1,4-diene	See γ-terpinene			
p-Mentha-1,4(8)-diene	See terpinolene			
p-Mentha-1,5-diene	See α-phellandrene			
p-Mentha-1,8-diene	See limonene			
Menthadienol	*p*-Mentha-1,8(10)-dien-9-ol	—	FDA§172.515	Synthetic flavor
p-Mentha-1,8-dien-7-ol	Perillyl alcohol	2664	FDA §172.515	Synthetic flavor

p-Mentha-1,8(10)-dien-9-ol	See menthadienol			
p-Mentha-6,8-dien-2-ol	See carveol			
6,8(9)-*p*-Menthadien-2-one	See carvone			
p-Mentha-6,8-dien-2-one	See carvone			
p-Mentha-1,4(8)-dien-3-one	1-Methyl-4-isopropylidene-1-cyclohexen-3-one	3560		See *Food Technol.*, February 1978
	Piperitenone	3560		See *Food Technol.*, February 1978
p-Menth-1,8-dien-7-yl-acetate	Dihydrocuminyl acetate	3561		See *Food Technol.*, February 1978
	4-Isopropenyl-1-cyclohexene carbinyl acetate	3561		See *Food Technol.*, February 1978
	Perillyl acetate	3561		See *Food Technol.*, February 1978
Menthadienyl acetate	*p*-Mentha-1,8(10)-dien-9-yl acetate	—	FDA §172.515	Synthetic flavor
1,8-*p*-Menthadien-7-yl acetate	See perillyl acetate			
p-Mentha-1,8-dien-7-yl acetate	See perillyl acetate			
p-Mentha-1,8(10)-dien-9-yl acetate	See menthadienyl acetate			
p-Mentha-6,8-dien-2-yl acetate	See carvyl acetate			
p-Mentha-6,8-dien-2-yl propionate	See carvyl propionate			
3-*p*-Menthanol	See menthol			
p-Menthan-2-ol	Carvomenthol	3562		See *Food Technol.*, February 1978
	Hexahydrocarvacrol	3562		See *Food Technol.*, February 1978
	3-Isopropyl-6-methyl-cyclohexanol	3562		See *Food Technol.*, February 1978
	1-Methyl-4-isopropyl-2-cyclohexanol	3562		See *Food Technol.*, February 1978
p-Menthan-2-one	Carvomenthone; tetrahydrocarvone	3176		10 ppm max in beverages, frozen desserts, and candies
p-Menthan-3-one	See menthone			
cis-*p*-Menthan-3-one	See *d,l*-isomenthone			
p-Mentha-8-thiol-3-one	8-Mercapto-*p*-menthane-3-one	3177		
p-Menth-1,4(8)-diene	See terpinolene			
p-Menth-1-ene-9-al		3178		
p-Menth-1-en-3-ol	Piperitol	3179		
p-Menth-1-en-8-ol	See α-terpineol			
p-Menth-3-en-1-ol		—	FDA §172.515	Synthetic flavor/ adjuvant

REGULATORY STATUS OF DIRECT FOOD ADDITIVES (continued)

Material	Alternate Name [Source]	FEMA No.	Regulation	Limitations
p-Menth-8-en-1-ol	See β-terpineol			
p-Menth-8-en-3-ol	See isopulegol			
p-8(9)-Menthen-3-ol	See isopulegol			
1-*p*-Menthen-4-ol	See 4-carvomenthenol			
1-*p*-Menthen-8-ol	See α-terpineol			
8-*p*-Menthen-2-ol	See dihydrocarveol			
1-*p*-Menthen-9-yl acetate	9-Acetoxy-1-*p*-menthene	3566		See *Food Technol.*, February 1978
p-Menth-1-en-8-ol butyrate	See terpinyl butyrate			
p-Menth-1-en-8-ol propionate	See terpinyl propionate			
p-Menth-1-en-3-one	See *d*-piperitone			
p-Menth-4(8)-en-3-one	See pulegone			
p-Menth-8-en-3-one	See isopulegone			
δ-4(8)-*p*-Menthen-3-one	See pulegone			
δ-8(9)-*p*-Menthen-3-one	See isopulegone			
p-Menth-1-en-8-yl acetate	See terpinyl acetate			
p-Menth-1-en-9-yl acetate	See 1-*p*-menthen-9-yl acetate			
p-Menth-3-en-1-ol	4-Isopropyl-1-methyl-3-cyclohexen-1-ol	3563		See *Food Technol.*, February 1978
	1-Methyl-4-isopropyl-3-cyclohexen-1-ol	3563		See *Food Technol.*, February 1978
	1-Terpineol	3563		See *Food Technol.*, February 1978
p-Menth-8-en-1-ol	4-Isopropenyl-1-methyl-1-cyclohexanol	3564		See *Food Technol.*, February 1978
	1-Methyl-4-isopropenylcyclohexan-1-ol	3564		See *Food Technol.*, February 1978
	β-Terpineol	3564		See *Food Technol.*, February 1978
p-Menth-8-en-2-one	*cis*-Dihydrocarvone	3565		See *Food Technol.*, February 1978
	3-Isopropenyl-6-methylcyclohexanone	3565		See *Food Technol.*, February 1978
p-Menth-8-en-3-yl acetate	See isopulegyl acetate			
p-Menth-8(9)-en-2-yl acetate	See dihydrocarvyl acetate			

1-*p*-Menthen-9-yl acetate		—	FDA §172.515	Synthetic flavor
8-*p*-Menthen-2-yl acetate	See dihydrocarvyl acetate			
1-*p*-Menthen-9-yl acetate	9-acetoxy-1-*p*- menthene	3566		See *Food Technol.*, February 1978
p-Menth-1-en-8-yl anthranilate	See terpinyl anthranilate			
p-Menth-1-en-8-yl butyrate	See terpinyl butyrate			
p-Menth-1-en-8-yl cinnamate	See terpinyl cinnamate			
p-Menth-1-en-8-yl formate	See terpinyl formate			
p-Menth-1-en-8-yl isobutyrate	See terpinyl isobutyrate			
p-Menth-1-en-8-yl isopentanoate	See terpinyl isovalerate			
p-Menth-1-en-8-yl isovalerate	See terpinyl isovalerate			
p-Menth-1-en-8-yl 3-methylbutanoate	See terpinyl isovalerate			
p-Menth-1-en-8-yl 3-methylbutyrate	See terpinyl isovalerate			
p-Menth-1-en-8-yl 3-phenylpropanoate	See terpinyl cinnamate			
p-Menth-1-en-8-yl propanoate	See terpinyl propanoate			
p-Menth-en-8-yl propionate	See terpinyl propionate			
Menthofuran	See 4,5,6,7-tetrahydro-3,6-dimethylbenzofuran			
Menthol	5-Methyl-2-*iso*-propylcyclohexanol	2665	FDA §172.515	Synthetic flavor
l-Menthol	See menthol			
d-neo-Menthol	See *d*-neomenthol			
Menthol acetate	See menthyl acetate			
Menthol isovalerate	See menthyl isovalerate			
Menthone	*p*-Menthan-3-one	2667	FDA §172.515	Synthetic flavor
Menthyl acetate	*p*-Menth-3-yl acetate	2668	FDA §172.515	Synthetic flavor
p-Menth-3-yl acetate	See menthyl acetate			
Menthyl isopentanoate	See menthyl isovalerate			
Menthyl isovalerate	*p*-Menth-3-yl isovalerate	2669	FDA §172.515	Synthetic flavor
p-Menth-3-yl isovalerate	See methyl isovalerate			
Menthyl isovalerianate	See menthyl isovalerate			

REGULATORY STATUS OF DIRECT FOOD ADDITIVES (continued)

Material	Alternate Name [Source]	FEMA No.	Regulation	Limitations
Menthyl 3-methylbutanoate	See menthyl isovalerate			
β-Mercaptoalanine	See *l*-cysteine			
2-Mercapto-3-butanol		3502		
3-Mercapto-2-butanol	See 2-mercapto-3-butanol			
3-Mercapto-2-butanone		3298		See *Food Technol.*, May 1972
4-Mercapto-2-butanone	See 2-keto-4-butanethiol			
8-Mercapto-*p*-menthane-3-one	See *p*-Mentha-8-thiol-3-one			
Mercaptomethane	See methyl mercaptan			
3-((2-mercapto-1-methyl-propyl)thio)-2-butanol	See α-methyl-β-hydroxypropyli α-methyl-β-mercaptopropyl sulfide			
2-Mercaptomethylpyrazine	Pyrazinemethanethiol	3299		See *Food Technol., May 1972*
2-Mercaptonaphthalene	See 2-naphthalenthiol			
3-Mercapto-2-pentanone		3300		See *Food Technol.*, May 1972
2-, 3-, or 10-Mercaptopinane		3503		
α-Mercaptopropanoic acid	See 2-mercaptopropionic acid			
2-Mercaptopropionic acid	Thiolactic acid	3180		
2-Mercaptopropionic acid, ethyl ester	See ethyl 2-mercaptopropionate			
α-Mercaptotoluene	See benzyl mercaptan			
Mesityl oxide	See 4-methyl-3-penten-2-one			
Methallyl caproate		—		
Methanethiol	See methyl mercaptan			
Methanethiol *n*-butyrate	See methyl thiobutyrate			
Methanoic acid	See formic acid			
Methialdol	See 2,8-dithianon-4-en-4-carboxaldehyde			
Methional	See 3-Methylthiopropionaldehyde			
D,L-Methionine		3301		See *Food Technol.*, May 1972
Methionol	See 3-(methylthio)propanol			
β-Methiopropionaldehyde	See 2-Methylthiopropionaldehyde			
4′-Methoxyacetophenone	See acetanisole			

p-Methoxyacetophenone	See acetanisole			
3-Methoxy-4-acetoxybenzaldehyde	See vanillin acetate			
p-Methoxyallylbenzene	See estragole			
2-Methoxy-4-allylphenol	See eugenol			
o-Methoxybenzaldehyde			FDA §172.515	Synthetic flavor
p-Methoxybenzaldehyde	*p*-Anisaldehyde	2670	FDA §172.515	Synthetic flavor
Methoxybenzene	See anisole			
p-Methoxybenzyl acetate	See anisyl acetate			
4-Methoxybenzyl acetate	See anisyl acetate			
p-Methoxybenzyl acetone	See 4-(*p*-methoxyphenyl)-2-butanone			
p-Methoxybenzyl alcohol	See anisyl alcohol			
p-Methoxybenzyl alcohol, formate	See anisyl formate			
p-Methoxybenzyl butyrate	See anisyl butyrate			
p-Methoxybenzyl formate	See anisyl formate			
4-Methoxybenzyl formate	See anisyl formate			
p-Methoxybenzyl methanoate	See anisyl formate			
p-Methoxybenzyl phenylacetate	See anisyl phenylacetate			
p-Methoxybenzyl propionate	See anisyl propionate			
2-Methoxy-3-*sec*-butylpyrazine	See 2-methoxy-3-(1-methylpropyl) pyrazine			
o-Methoxycinnamaldehyde		3181	FDA §172.51 5	Synthetic flavor
2-Methoxy-*p*-cresol	See 2-methoxy-4-methylphenol			
3-Methoxy-*p*-cymene	See 1-methyl-3-methoxy-4-isopropylbenzene			
2-Methoxy-1,4-diazine	See methoxypyrazine			
2-Methoxy-4-ethylphenol	See 4-ethyl guaiacol			
2,5, or 6-Methoxy-3-ethylpyrazine	See 2-ethyl (or methyl)-(3,5, or 6)-methoxypyrazine			
3-Methoxy-4-hydroxybenzaldehyde	See vanillin			
3-Methoxy-4-hydroxybenzylacetone	See zingerone			

REGULATORY STATUS OF DIRECT FOOD ADDITIVES (continued)

Material	Alternate Name [Source]	FEMA No.	Regulation	Limitations
3-Methoxy-4-hydroxy-1-propen-1-yl benzene	See iso-eugenol			
3-Methoxy-4-hydroxytoluene	See 2-methoxy-4-methylphenol			
2-Methoxy-3(5 and 6)-isopropylpyrazine		3358		
o-Methoxy methyl benzoate	See methyl *o*-methoxybenzoate			
p-Methoxy-α-methyl cinnamaldehyde		3182		
2-Methoxy-4-methylphenol	2-Methoxy-*p*-cresol; 4-methylguaiacol	2671	FDA §172.515	Synthetic flavor
2-Methoxy-3-(1-methylpropyl)pyrazine		3433		
2-, 5-, or 6-Methoxy-3-methylpyrazine (mixture of isomers)		3183		10 ppm max in finished food
p-Methoxy-β-methylstyrene	See anethole			
o-Methoxyphenol	See guaiacol			
o-Methoxyphenyl acetate	See guaiacyl phenylacetate			
O-Methoxyphenyl acetate	See guaiacyl acetate			
P-Methoxyphenylacetone	See 1-(*p*-methoxyphenyl)-2-propanone			
4-Methoxyphenyl acetone	See 1-(*p*-methoxyphenyl)-2-propanone			
β-(*O*-Methoxyphenyl) acrolein	See *o*-methoxycinnamaldehyde			
4-(*p*-Methoxyphenyl)-2-butanone	Anisyl acetone	2672	FDA §172.515	Synthetic flavor
1-(4-Methoxyphenyl)-4-methyl-1-penten-3-one		—	FDA §172.515	Synthetic flavor
2-(4-Methoxyphenyl)-4-methyl-1-penten-3-one	Methoxystyryl isopropyl ketone		FDA §172.515	Synthetic flavor/adjuvant
3-(*p*-Methoxyphenyl)-2-methyl-2-propenal	See *p*-methoxy-α-methyl cinnamaldehyde			
1-(*p*-Methoxyphenyl)-1-penten-3-one	Ethone; α-methylanisylideneacetone	2673	FDA §172.515	Synthetic flavor

1-(*p*-Methoxyphenyl)-2-propanone	Anisic ketone; anisyl methyl ketone	2674	FDA §172.515	Synthetic flavor
3-(*O*-Methoxyphenyl)-2-propenal	See *o*-methoxycinnamaldehyde			
p-Methoxy-*α*-phenyl propene	See anethole			
p-Methoxypropenylbenzene	See anethole			
1-Methoxy-4-propenylbenzene	See anethole			
1-Methoxy-4-(2-propen-1-yl)benzene	See estragole			
4-Methoxy-1-propenylbenzene	See anethole			
2-Methoxy-4-propenylphenol	See iso-eugenol			
2-Methoxy-4-(2-propen-1-yl)phenol	See eugenol			
2-Methoxy-4-propenylphenyl acetate	See iso-eugenyl acetate			
2-Methoxy-4-(2-propen-1-yl)phenyl acetate	See eugenyl acetate			
2-Methoxy-4-propenylphenyl benzyl ether	See isoeugenyl benzyl ether			
2-Methoxy-4-propenylphenyl formate	See isoeugenyl formate			
2-Methoxy-4-(1-propen-1-yl)phenyl formate	See isoeugenyl formate			
2-Methoxy-4-propenylphenyl phenyl acetate	See isoeugenyl phenyl acetate			
2-Methoxy-4-(1-propen-1-yl)phenyl phenylacetate	See isoeugenyl phenylacetate			
1-Methoxy-4-propylbenzene	See *p*-propylanisole			
Methoxypyrazine	2-Methoxy-1,4-diazine	3302		See *Food Technol.*, May 1972
2-Methoxystyrene	See *o*-vinylanisole			
P-Methoxystyryl ethyl ketone	See 1-(*p*-methoxyphenyl)-1-penten-3-one			

REGULATORY STATUS OF DIRECT FOOD ADDITIVES (continued)

Material	Alternatate Name [Source]	FEMA No.	Regulation	Limitations
Methoxystyryl isopropyl ketone	See 1-(4-methoxyphenyl)-4-methyl-1-penten-3-one			
o-Methoxytoluene	See *o*-methylanisole			
p-Methoxytoluene	See *p*-methylanisole			
2-Methoxytoluene	See *o*-methylanisole			
4-Methoxytoluene	See *p*-methylanisole			
1-Methoxy-2-vinylbenzene	See *o*-vinylanisole			
2-Methoxy-4-vinylphenol	*p*-Vinylguaiacol	2675	FDA §172.515	Synthetic flavor
Methylacetaldehyde	See propionaldehyde			
Methyl acetate		2676	FDA§172.515	Synthetic flavor
Methylacetic acid	See propionic acid			
p-Methylacetophenone	See 4′-methylacetophenone			
4-Methylacetophenone	*p*-methylacetophenone; methyl *p*-tolyl ketone	2677	FDA §172.515	Synthetic flavor
1-Methyl-4-acetyl benzene	See 4′-methylacetophenone			
1-Methyl-2-acetylpyrrole		3184		
Methyl alcohol			FDA §8.305	Solvent for annatto extract; residues, 50 ppm max
2-Methylallyl butanoate	See 2-methylallyl butyrate			
2-Methylallyl butyrate	2-methyl-2-propen-1-yl butyrate	2678	FDA §172.515	Synthetic flavor
Methyl allyl disulfide	See allyl methyl disulfide			
Methyl allyl trisulfide	See allyl methyl trisulfide			
Methyl 2-aminobenzoate	See methyl anthranilate			
2-Methylamino methylbenzoate	See methyl *N*-methylanthranilate			
Methylamylacetic acid	See 2-methylheptanoic acid			
Methyl *n*-amyl carbinol	See 2-heptanol			
Methyl amyl ketone	See 2-heptanone			
Methyl anisate		2679	FDA §172.515	Synthetic flavor
Methyl *o*-anisate	See methyl *o*-methoxybenzoate			
o-Methylanisole	*o*-cresyl methyl ether	2680	FDA §172.515	Synthetic flavor
p-Methylanisole	*p*-cresyl methyl ether; *p*-methoxytoluene	2681	FDA §172.515	Synthetic flavor

α-Methyl anisylidene acetone	See 1-(*p*-methoxyphenyl)-1-penten-3-one			
Methyl anthranilate	methyl-2-aminobenzoate	2682	FDA §182.1	GRAS, synthetic flavor and adjuvant
Methylated silica		3185		
Methylben	See methyl *p*-hydroxybenzoate			
α-Methyl-α-benzalacetone	See 3-methyl-4-phenyl-3-butene-2-one			
Methylbenzaldehyde (mixed 2, 3, 4)	See tolualdehydes (mixed *o, m, p*)			
Methyl benzenecarboxylate	See methyl benzoate			
Methyl benzoate	oil of niobe	2683	FDA §172.515	Synthetic flavor
6-Methyl-2*H*-1-benzopyran-2-one	See 6-methylcoumarin			
6-Methylbenzopyrone	See 6-methylcoumarin			
Methylbenzyl acetate (mixed *o*-, *m*-, *p*-)		—	FDA §172.515	Synthetic flavor/adjuvant
α-Methylbenzyl acetate	styralyl acetate	2684	FDA§172.515	Synthetic flavor
p-Methylbenzyl acetone	See 4-(*p*-tolyl)-2-butanone			
α-Methylbenzyl alcohol	styralyl alcohol	2685	FDA §172.515	Synthetic flavor
α-Methylbenzyl butyrate	styralyl butyrate	2686	FDA §172.515	Synthetic flavor
Methylbenzyl disulfide		3504		
α-Methylbenzyl formate	styralyl formate	2688	FDA §172.515	Synthetic flavor
1-Methyl-1-benzylidene acetone	See 3-methyl-4-phenyl-3-butene-2-one			
α-Methylbenzyl isobutyrate	styralyl isobutyrate	2687	FDA §172.515	Synthetic flavor
α-Methylbenzyl methanoate	See α-methylbenzyl formate			
α-Methylbenzyl 2-methylpropanoate	See α-methylbenzyl isobutyrate			
α-Methylbenzyl propionate	styralyl propionate	2689	FDA §172.515	Synthetic flavor
4-Methylbiphenyl	*p*-methyldiphenyl	3186		
2-Methylbutanal	See 2-methylbutyraldehyde			
3-Methylbutanal	See 3-methylbutyraldehyde			
2-Methylbutane-1-carboxylic acid	See 3-methylpentanoic acid			
3-Methylbutane-1-carboxylic acid	See 4-methylpentanoic acid			
Methyl butanedioate	See dimethyl succinate			
2-Methyl-1-butanethiol		3303		See *Food Technol.*, May 1972
3-Methyl-2-butanethiol		3304		See *Food Technol.*, May 1972

REGULATORY STATUS OF DIRECT FOOD ADDITIVES (continued)

Material	Alternatate Name [Source]	FEMA No.	Regulation	Limitations
Methyl butanoate	See methyl butyrate			
3-Methylbutanoic acid	See isovaleric acid			
3-Methyl-1-butanol	See isoamyl alcohol			
α-Methylbutanol	See 2-pentanol			
2-Methyl-2-butenal		3407		
3-Methyl-3-buten-2-one		—		
3-Methylbutyl acetate	See isoamyl acetate			
β-Methyl butyl acetate	See isoamyl acetate			
3-Methylbutyl acetoacetate	See isoamyl acetoacetate			
3-Methylbutylamine	See isopentylamine			
3-Methylbutyl benzoate	See isoamyl benzoate			
3-Methylbutyl butanoate	See isoamyl butyrate			
3-Methylbutyl *o*-hydroxybenzoate	See isoamyl salicylate			
3-Methylbutyl isobutyrate	See isoamyl isobutyrate			
2-Methylbutyl isopentanoate	See 2-methylbutyl 3-methylbutanoate			
2-Methylbutyl isopentanoate	See 2-methylbutyl isovalerate			
2-Methylbutyl isovalerate		—	FDA §172.515	Synthetic flavor/adjuvant
2-Methylbutyl isovalerate	See 2-methylbutyl 3-methylbutanoate			
2-Methylbutyl isovalerianate	See 2-methylbutyl 3-methylbutanoate			
2-Methylbutyl isovalerianate	See 2-methylbutyl isovalerate			

3-Methylbutyl β-ketobutyrate	See isoamyl acetoacetate			
2-Methylbutyl 2-methylbutanoate	See 2-methylbutyl 2-methylbutyrate			
2-Methylbutyl 3-methylbutanoate		3506	FDA §172.515	Synthetic flavor
3-Methylbutyl 2-methylbutanoate	Isoamyl 2-methylbutyrate	3505	FDA§172.515	Synthetic flavor
3-Methylbutyl 3-methylbutanoate	See isoamyl isovalerate			
2-Methylbutyl 2-methylbutyrate		3359		
3-Methylbutyl 3-methylbutyrate	See isoamyl isovalerate			
3-Methylbutyl 2-methyl propanoate	isoamyl isobutyrate	3507	FDA §172.515	Synthetic flavor
3-Methylbutyl nonanoate	See isoamyl nonanoate			
3-Methylbutyl octanoate	See isoamyl octanoate			
3-Methylbutyl 3-oxobutanoate	See isoamyl acetoacetate			
3-Methylbutyl pelargonate	See isoamyl nonanoate			
3-Methylbutyl pentanoate	See isoamyl valerate			
3-Methylbutyl phenylacetate	See isoamyl phenylacetate			
Methyl *p-tert*-butylphenylacetate		2690	FDA §172.515	Synthetic flavor

REGULATORY STATUS OF DIRECT FOOD ADDITIVES (continued)

Material	Alternate Name [Source]	FEMA No.	Regulation	Limitations
3-Methylbutyl propanoate	See isoamyl propionate			
3-Methylbutyl propionate	See isoamyl propionate			
3-Methylbutyl salicylate	See isoamyl salicylate			
2-Methylbutyraldehyde	methylethylacetaldehyde	2691	FDA §172.515	Synthetic flavor
3-Methylbutyraldehyde	isovaleraldehyde	2692	FDA §172.515	Synthetic flavor
Methylbutyrate		2693	FDA §172.515	Synthetic flavor
2-Methylbutyric acid		2695	FDA §172.515	Synthetic flavor
3-Methylbutyric acid	See isovaleric acid			
β-Methylbutyric acid	See isovaleric acid			
3-Methylbutyrolactone	See γ-valerolactone			
γ-Methyl-γ-butyralactone	See γ-valerolactone			
Methyl caproate	See methyl hexanoate			
Methyl caprylate	See methyl octanoate			
Methyl carbinol	See ethyl alcohol			
Methylcatechol	See guaiacol			
O-Methylcatechol	See guaiacol			
O-Methylcatechol acetate	See guaiacyl phenylacetate			
Methylcellulose	cellulose methyl ether	2696	FDA §182.1	GRAS, miscellaneous and/or general-purpose food additive; product specifications apply
			FDA §25.2 and §25.3	In French and salad dressings
			MID	Carrier and extender and stabilizer of meat and vegetable patties, at 0.15%
Methyl chavicol	See estragole			
p-Methylcinnamaldehyde		—	FDA §172.515	Synthetic flavor/adjuvant
α-Methylcinnamaldehyde		2697	FDA §172.515	Synthetic flavor
Methyl cinnamate		2698	FDA §172.515	Synthetic flavor
α-Methylcinnamic aldehyde	See α-methylcinnamaldehyde			
Methyl citronellate	See methyl 3,7-dimethyl-6-octenoate			
6-Methylcoumarin	5-Methyl-2-hydroxyphenylpropenoic acid lactone	2699		

6-Methyl-*cis*-*o*-coumarinic lactone	See 6-methylcoumarin			
O-Methyl-*p*-cresol	See *p*-methylanisole			
2-Methylcrotonaldehyde	See 2-methyl-2-butenal			
3-Methylcrotonic acid	3,3-Dimethylacrylic acid; β,β-dimethylacrylic acid; senecioic acid	3187		
p-Methylcumene	See *p*-cymene			
1-Methyl-1,3-cyclohexadiene		—	FDA §172.515	Synthetic flavor
2-Methyl-1,3-cyclohexadiene			FDA §172.515	Synthetic flavor/adjuvant
1-Methyl-2,3-cyclohexadione		3305		See *Food Technol.*, May 1972
3-Methyl-2-cyclohexen-1-one		3360		
3-Methylcyclopentadecanone	See 3-methyl-1-cyclopentadecanone			
3-Methyl-1-cyclopentadecanone		3434		
3-Methylcyclopentane-1,2-dione	See methylcyclopentenolone			
Methylcyclopentenolone	3-methylcyclopentane-1,2-dione	2700	FDA §172.515	Synthetic flavor
1-Methyl-1-cyclopenten-3-one		3435		
3-Methyl-2-cyclopenten-1-one	See 1-methyl-1-cyclopenten-3-one			
Methyl decine carbonate	See methyl 2-undecynoate			
Methyl decyne carbonate	See methyl-2-undecynoate			
2-Methyl-1,4-diazine	See 2-methylpyrazine			
Methyl dihydroabietate	See methyl ester of rosin, partially hydrogenated			
Methyl dihydrocinnamate	See methyl 3-phenylpropionate			
5*H*-5-Methyl-6,7-dihydrocyclopenta [*b*] pyrazine		3306		See *Food Technol.*, May 1972
Methyl dihydrojasmonate		3408		
2-Methyl-4,5(dihydro)-3(2*H*)-thiophenone	See 2-methyltetrahydrothiophen-3-one			

REGULATORY STATUS OF DIRECT FOOD ADDITIVES (continued)

Material	Alternate Name [Source]	FEMA No.	Regulation	Limitations
Methyl 3,7-dimethyl-6-octenoate		3361		
Methyl 2,4-dimethylphenyl ketone	See 2,4-dimethylacetophenone			
p-Methyldiphenyl	See 4-methylbiphenyl			
Methyl disulfide	See dimethyl disulfide			
Methyldithiopropane	See methyl propyl disulfide			
Methyldithio-1-propene	See methyl propenyl disulfide and methyl 1-propenyl disulfide			
Methyl dodecanoate	See methyl laurate			
Methyl dodecylate	See methyl laurate			
Methylene chloride	Dichloromethane		FDA §8.390	Diluent in inks for marking fruits and vegetables, with no residue
			FDA §8.305	Solvent for annatto extract residues, 30 ppm max; total residues of chlorinated solvents, 30 ppm max
4-(2-Methylene-6,6-dimethylcyclohexyl)-3-buten-2-one	See γ-ionone			
3,4-Methylenedioxybenzaldehyde	See piperonal			
3,4-Methylenedioxybenzyl acetate	See piperonyl acetate			
3,4-Methylenedioxybenzyl 2-methylpropanoate	See piperonyl isobutyrate			
4-(3,4-Methylenedioxyphenyl)-2-butanone		2701		
2-Methylene-6,10,10-trimethylbicyclo (7.2.0)undec-5-ene	See β-caryophylalene			
Methyl ester of rosin, partially hydrogenated	methyl dihydroabietate		FDA §172.515	Synthetic flavor

Methyl ethanoate	See methyl acetate			
Methyl ethyl acetaldehyde	See 2-methylbutyraldehyde			
Methylethylacetic acid	See 2-methylbutyric acid			
α-Methyl-β-ethylacrolein	See 2-methyl-2-pentenal			
1-Methylethyl benzoate	See isopropyl benzoate			
Methyl ethyl cellulose			FDA §121.1112	Aerating, emulsifying, and foaming agent; product specifications apply
Methyl ethyl ketone	See 2-butanone			
1-Methylethyl 3-phenylpropenoate	See isopropyl cinnamate			
2-Methyl-5-ethylpyridine	See 5-ethyl-2-methylpridine			
2-Methyl-5-ethylpyrazine	See 2-ethyl-5-methyl pyrazine			
Methyl eugenol	See eugenyl methyl ether			
O-Methyl eugenol	See eugenyl methyl ether			
Methyl eugenol ether	See eugenyl methyl ether			
Methyl eugenyl ether	See eugenyl methyl ether			
Methylexaltone	See 3-methyl-1-cyclopentadecanone			
5-Methyl-2(3 *H*)-furanone	See 4-hydroxy-3-pentenoic acid lactone			
2-Methyl-3-furanthiol		3188		
Methyl furfuracylate		2267		
5-Methylfurfural		2702		
Methyl furfuryl disulfide		3362		
2-Methyl-3-, 5-, or 6-furfuryl thiopyrazine (mixture of isomers)		3189		10 ppm max in finished food
Methyl 2-furoate	methyl pyromucate	2703		
2-Methyl-3-furylacrolein	α-methyl furylacrolein	2704		
2-Methyl-3-furylacrolein	See furfurylidene-2-propanal			
α-Methyl furylacrolein	See 2-methyl-3-furylacrolein			
Methyl 3-(2-furyl)acrylate	See methyl furfuracylate			
3-(5-Methyl-2-furyl)-butanal	3-(5-methyl-2-furyl)-butyraldehyde	3307		
3-(5-Methyl-2-furyl)-butyraldehyde	See 3-(5-methyl-2-furyl)-butanal			
2-Methyl-3-furyl disulfide	See bis(2-methyl-3-furyl)disulfide			
2-Methyl-3-furyl tetrasulfide	See bis(2-methyl-3-furyl)-tetrasulfide			

REGULATORY STATUS OF DIRECT FOOD ADDITIVES (continued)

Material	Alternate Name [Source]	FEMA No.	Regulation	Limitations
3-[(2-Methyl-3-furyl) thio]-2,6-dimethyl-4-heptanone	See 2,6-dimethyl-3-[(2-methyl-3-furyl) thio]-4-heptanone			
4[(2-Methyl-3-furyl) thio]-5-nonanone	1,3-dipropylacetonyl 2-methyl-3-furyl sulfide	3571		See *Food Technol.*, February 1978
Methyl glucoside — coconut oil ester			FDA §121.1151	Acid in crystallization of sucrose and dextrose; 320 ppm max im molasses as surface-active agent; product specifications apply
Methyl glycol	See propylene glycol			
Methyl glyoxal	See pyruvaldehyde			
4-Methylguaiacol	See 2-methoxy-4-methylphenol			
2-Methylhendecanal	See 2-methyl undecenal			
Methylheptadienone	See 6-methyl-3,5-heptadien-2-one			
6-Methyl-3,5-heptadien-2-one		3363		
Methyl heptanoate		2705	FDA §172.515	Synthetic flavor
2-Methylheptanoic acid	2-methyloenanthic acid	2706	FDA §172.515	Synthetic flavor
2-Methyl-2-hepten-6-one	See 6-methyl-5-hepten-2-one			
6-Methyl-5-hepten-2-one		2707	FDA §172.515	Synthetic flavor
Methyl heptine carbonate	See methyl 2-octynoate			
Methyl heptoate	See methyl heptanoate			
Methyl heptylate	See methyl heptanoate			
Methyl *n*-heptyl carbinol	See 2-nonanol			
Methyl heptyl ketone	See 2-nonanone			
Methyl heptyne carbonate	See methyl 2-octynoate			
5-Methyl-2,3-hexanedione		3190		
Methyl hexanoate		2708	FDA §172.515	Synthetic flavor
2-Methylhexanoic acid		3191		
Methyl 2-hexenoate		2709	FDA §172.515	Synthetic flavor
Methyl 3-hexenoate		3364		
5-Methyl-3-hexen-2-one		3409		
5-Methyl-5-hexen-2-one		3365		
Methyl hexyl acetaldehyde	See 2-methyl octanal			

Methyl hexyl carbinol	See 2-octanol	
Methyl hexyl ketone	See 2-octanone	
p-Methylhydratropaldehyde	See 2-(*p*-tolyl)propionaldehyde	
p-Methylhydratropicaldehyde	See 2-(*p*-tolyl)propionaldehyde	
Methyl hydrocinnamate	See methyl-3-phenylpropionate	
1-Methyl-3-hydroxybenzene	See *m*-cresol	
1-Methyl-4-hydroxybenzene	See *p*-cresol	
Methyl *p*-hydroxybenzoate	See methylparaben	2710
Methyl 2-hydroxybenzoate	See methyl salicylate	
4-Methyl-4-hydroxybutanoic acid lactone	See γ-valerolactone	
Methyl 3-hydroxycaproate	See methyl 3-hydroxyhexanoate	
Methyl β-hydroxycaproate	See methyl 3-hydroxyhexanoate	
4-Methyl-5-(β-hydroxyethyl)-thiazole	See 4-methyl-5-thiazole-ethanol	
Methyl 3-hydroxyhexanoate		3508
Methyl β-hydroxyhexanoate	See methyl 3-hydroxyhexanoate	
1-Methyl-3-hydroxy-4-isopropylbenzene	See thymol	
5-Methyl-5-hydroxypentanoic acid lactone	See δ-hexalactone	
5-Methyl-2-hydroxyphenylpropenoic acid lactone	See 6-methylcoumarin	
Methyl (2-hydroxyphenyl-sulfide	See *o*-(methylthio)-phenol	
α-Methyl-β-hydroxypropyl α-methyl-β-mercaptopropyl sulfide		3509
Methyl 3-hydroxypropyl sulfide	See 3-(methylthio)propanol	
3-Methylindole	See skatole	
β-Methylindole	See skatole	

REGULATORY STATUS OF DIRECT FOOD ADDITIVES (continued)

Material	Alternate Name [Source]	FEMA No.	Regulation	Limitations
6-Methylionone	See α-irone			
β-Methylionone	See methyl β-ionone			
γ-Methylionone	See α-iso-methylionone			
α-iso-Methylionone	See α-isomethylionone			
Methyl α-ionone	5-(2,6,6-trimethyl-2-cyclohexen-1-yl)-4-penten-3-one	2711	FDA §172.515	Synthetic flavor
Methyl β-ionone	5-(2,6,6-trimethyl-1-cyclohexen-1-yl)-4-penten-3-one	2712	FDA §172.515	Synthetic flavor
Methyl γ-ionone	See α-isomethylionone			
Methyl δ-ionone	5-(2,6,6-trimethyl-3-cyclohexen-1-yl)-4-penten-3-one	2713	FDA §172.515	Synthetic flavor
Methyl isobutenyl ketone	See 4-methyl-3-penten-2-one			
Methyl isobutylacetate	See methyl-4-methylvalerate			
Methyl isobutyl ketone	See 4-methyl-2-pentanone			
2-Methyl-3-isobutylpyrazine	See 2-isobutyl-3-methylpyrazine			
Methyl isobutyrate		2694	FDA §172.515	Synthetic flavor
Methyl isocaproate	See methyl 4-methylvalerate			
Methyl isoeugenol	See isoeugenyl methyl ether			
Methyl isopentanoate	See methyl isovalerate			
1-Methyl-4-isopropenylbenzene	See *p*-α-dimethylstyrene			
6-Methyl-3-isopropenylcyclohexanol	See dihydrocarveol			
1-Methyl-4-isopropenylcyclohexan-1-ol	See β-terpineol and *p*-menth-8-en-1-ol			
1-Methyl-4-isopropenylcyclohexan-3-ol	See isopulegol			
1-Methyl-4-isopropenyl-3-cyclohexanone	See isopulegone			
1-Methyl-4-isopropenylcyclohexan-2-yl acetate	See dihydrocarvyl acetate			

1-Methyl-4-isopropenylcyclohexan-3-yl acetate	See isopulegyl acetate
1-Methyl-4-isopropenyl-1-cyclohexene	See *d*-limonene
1-Methyl-4-isopropenyl-6-cyclohexen-2-ol	See carveol
1-Methyl-4-isopropenyl-6-cyclohexen-2-one	See carvone
6-Methyl-3-isopropenylcyclohexyl acetate	See dihydrocarvyl acetate
1-Methyl-4-isopropylbenzene	See *p*-cymene
4-Methyl-1-isopropylbenzene	See *p*-cymene
1-Methyl-4-isopropyl-1,3-cyclohexadiene	See α-terpinene
1-Methyl-4-isopropyl-1,3-cyclohexadiene	See *p*-Mentha-1,3-diene
1-Methyl-4-isopropyl-1,4-cyclohexadiene	See γ-terpinene and *p*-mentha,1,4-diene
1-Methyl-4-isopropyl-1,5-cyclohexadiene	See α-phellandrene
2-Methyl-5-isopropyl-1,3-cyclohexadiene	See α-phellandrene
1-Methyl-4-isopropyl-2-cyclohexanol	See *p*-methan-2-ol
5-Methyl-2-isopropylcyclohexanol	See menthol
1-Methyl-4-isopropylcyclohexan-3-ol	See menthol
1-Methyl-4-isopropyl-3-cyclohexanone	See *d,l*-isomenthone
1-Methyl-4-isopropyl-1-cyclohexene-4-ol	See 4-carvomenthenol
1-Methyl-4-isopropyl-3-cyclohexen-1-ol	See *p*-menth-3-en-1-ol
1-Methyl-4-isopropyl-1-cyclohexen-8-ol	See α-terpineol

REGULATORY STATUS OF DIRECT FOOD ADDITIVES (continued)

Material	Alternate Name [Source]	FEMA No.	Regulation	Limitations
1-Methyl-4-isopropyl-1-cyclohexen-3-one	See *d*-piperitone			
Methyl isopropyl diketone	See 4-methyl-2,3-pentanedione			
5-Methyl-2-isopropylhexahydrophenol	See menthol			
α-Methyl-*p*-isopropylhydrocinnamaldehyde	See 2-methyl-3-(*p*-isopropylphenyl)propionaldehyde			
1-Methyl-4-isopropylidene-3-cyclohexanone	See pulegone			
1-Methyl-4-isopropylidene-1-cyclohexene	See terpinolene			
1-Methyl-4-isopropylidene-1-cyclohexen-3-one	See piperitenone			
2-Methyl-5-isopropylphenol	See carvacrol			
3-Methyl-6-isopropylphenol	See thymol			
5-Methyl-2-isopropylphenol	See thymol			
Methyl *p*-isopropylphenyl ketone	See *p*-isopropylacetophenone			
2-Methyl-3-(*p*-isopropylphenyl)propionaldehyde	cyclamen aldehyde; α-methyl-*p*-isopropylhydrocinnamaldehyde	2743	FDA §172.515	Synthetic flavor
α-Methyl-*p*-isopropylphenylpropylaldehyde	See 2-methyl-3-*p*-isopropylphenyl)propionaldehyde			
2-Methyl-5-isopropylpyrazine	See 5-isopropyl-2-methylpyrazine	—		
Methyl isovalerate		2753	FDA §172.515	Synthetic flavor
Methyl isovalerianate	See methyl isovalerate			
Methyl jasmonate		3410		
Methyl laurate	methyl dodecanoate	2715	FDA §172.515	Synthetic flavor
Methyl linoleate (48%)-		3411		

methyl linolenate (52%) mixture				
Methyl mercaptan	Methanethiol	2716	FDA §172.515	Synthetic flavor
(Methylmercapto) acetaldehyde	See 2-methylthioacetaldehyde			
γ-(Methylmercapto)-butyraldehyde	See 4-(methylthio)butanal			
3-(Methylmercapto)-1-hexanol	See 3-(methylthio)-1-hexanol			
2-Methylmercaptophenol	See *o*-(methylthio)-phenol			
Methylmercaptopropionaldehyde	See 3-(methylthio)propionaldehyde			
3-(Methylmercaptopropionaldehyde	See 3-(methylthio)propionaldehyde			
β-(Methylmercapto)propionaldehyde	See 3-(methylthio)propionaldehyde			
γ-(Methylmercapto)propyl alcohol	See 3-(methylthio)propanol			
3-Methylmercaptopropyl isothiocyanate	See 3-methylthiopropyl isothiocyanate			
dl-Methylmethionine sulfonium chloride	See *dl*-(3-amino-3-caroxypropyl)dimethylsulfonium chloride			
Methyl β-methiopropionate	See methyl 3-methylthiopropionate			
Methyl *o*-methoxybenzoate	*o*-Methoxy methyl benzoate	2717	FDA §172.515	Synthetic flavor
Methyl *p*-methoxybenzoate	See methyl anisate			
1-Methyl-3-methoxy-4-isopropylbenzene		3436		
Methyl 4-methoxyphenyl ketone	See acetanisole			
2-Methyl (or ethyl)-(3,5, and 6)-methoxypyrazine	See 2-ethyl-(3,5, and 6)-methoxypyrazine (85%) plus 2-methyl-(3,5, and 6)-methoxypyrazine (13%)			
2-Methyl-5-methoxythiazole		3192		
Methyl *N*-methylanthranilate	Dimethyl anthranilate	2718	FDA §172.515	Synthetic flavor
Methyl 2-methylbutanoate	See methyl 2-methylbutyrate			

REGULATORY STATUS OF DIRECT FOOD ADDITIVES (continued)

Material	Alternatate Name [Source]	FEMA No.	Regulation	Limitations
Methyl 3-methylbutanoate	See methyl isovalerate			
Methyl-2-methylbutyrate		2719	FDA §172.515	Synthetic flavor/adjuvant
Methyl-3-methylbutyrate	See methyl isovalerate			
7-Methyl-3-methylene-1,6-octadiene	See myrcene			
5-Methyl-2-(1-methylethyl)phenol	See thymol			
Methyl γ-(methylmercapto)-butyrate	See methyl 4-(methylthio) butyrate			
Methyl β-(methylmercapto) propionate	See methyl 3-methylthiopropionate			
Methyl 4-methylpentanoate	See methyl 4-methylvalerate			
p-Methyl-α-methylphenyl acetaldehyde	See 2-(*p*-tolyl)-propionaldehyde			
Methyl 2-methylpropanoate	See methyl isobutyrate			
Methyl 4-(methylthiobutyr-ate		3412		
Methyl γ-(methylthio)bu-tyrate	See methyl 4-(methylthio)butyrate			
2-Methyl-5-(methylthio)furan		3366		
Methyl 3-methylthiopro-pionate	Methyl β-methyl mercaptopropionate; methyl β-methiopropionate	2720	FDA §172.515	Synthetic flavor
2-Methyl-3-, 5-, or 6-meth-ylthiopyrazine	See (methylthio)methylpyrazine (mix-ture of isomers)			
Methyl 4-methylvalerate	Methyl 4-methylpentanoate	2721	FDA §172.515	Synthetic flavor
Methyl myristate	Methyl tetradecanoate	2722	FDA §172.515	Synthetic flavor
1-Methylnaphthalene		3193		
α-Methylnaphthalene	See 1-methylnaphthalene			
Methyl β-naphthyl ether	See β-naphthyl methyl ether			
Methyl β-naphthyl ketone	2′-Acetonaphthone	2723	FDA §172.515	Synthetic flavor

Methyl nonanoate		2724	FDA §172.515	Synthetic flavor
Methyl 2-nonenoate	Methyl octyne carbonate	2725	FDA §172.515	Synthetic flavor
Methyl nonyl acetaldehyde	See 2-methylundecanal			
Methyl nonylate	See methyl nonanoate			
Methyl nonyl carbinol	See 2-undecanol			
Methyl nonylenate	See methyl 2-nonenoate			
Methyl nonyl ketone	See 2-undecanone			
Methyl 2-nonynoate	Methyl octyne carbonate	2726	FDA §172.515	Synthetic flavor
Methyl 9,12-octadecadienoate-methyl 9,12,-15-octadecatrienoate mixture	See methyl linoleate (48%)-methyl linolenate (52%) mixture			
2-Methyloctanal	Methyl hexyl acetaldehyde	2727	FDA §172.515	Synthetic flavor
Methyl octanoate		2728	FDA §172.515	Synthetic flavor
Methyl *cis*-4-octenoate		3367		
Methyl octylate	See methyl octanoate			
Methyl octyne carbonate	See methyl 2-nonynoate			
Methyl 2-octynoate	Methyl heptine carbonate	2729	FDA §172.515	Synthetic flavor
2-Methyloenanthic acid	See 2-methylheptanoic acid			
Methylol methyl amylketone	See 3-octanon-1-ol			
Methyl oxanone	See 4-(*p*-methoxyphenyl)-2-butanone			
Methylparaben	Methyl *p*-hydroxybenzoate	2710	FDA §182.1	GRAS, chemical preservative, up to
				0.1%
			FDA §172.515	Synthetic flavor
Methyl parasept	See methyl *p*-hydroxybenzoate			
Methyl pelargonate	See methyl nonanoate			
2-Methylpentanal		3413		
4-Methyl-2,3-pentanedione	acetyl isobutyryl	2730	FDA §172.515	Synthetic flavor
Methyl pentanoate	See methyl valerate			
2-Methylpentanoic acid	See 2-methylvaleric acid			
3-Methylpentanoic acid		3437		
4-Methylpentanoic acid		3463		
4-Methyl-2-pentanone	methyl isobutyl ketone	2731	FDA §172.515	Synthetic flavor
2-Methyl-2-pentenal		3194		
4-Methyl-2-pentenal		3510		
2-Methyl-2-pentenoic acid		3195		
2-Methyl-3-pentenoic acid		3464		
2-Methyl-4-pentenoic acid		3511		

REGULATORY STATUS OF DIRECT FOOD ADDITIVES (continued)

Material	Alternatate Name [Source]	FEMA No.	Regulation	Limitations
4-Methyl-3-penten-2-one		3368		
4-Methyl-4-penten-2-one		—		
3-Methyl-2-(2-pentenyl)-2-cyclopenten-1-one	jasmone; *cis*-jasmone	3196		
Methyl pentenyl ketone	See 3-hepten-2-one			
Methyl (2-pent-2-enyl-3-oxo-1-cyclopentyl) acetate	See methyl jasmonate			
Methyl (2-pentyl-3-oxo-1-cyclopentyl) acetate	See methyl dihydrojasmonate			
β-Methylphenethyl alcohol	hydratropyl alcohol	2732	FDA §172.515	Synthetic flavor
α-Methylphenethyl butyrate	1-phenyl-2-propyl butyrate	3197		
β-Methylphenethyl butyrate	See 2-phenylpropyl butyrate			
Methyl phenethyl ether		3198		
m-Methylphenol	See *m*-cresol			
o-Methylphenol	See *o*-cresol			
p-Methylphenol	See *p*-cresol			
3-Methylphenol	See *m*-cresol			
p-Methylphenylacetaldehyde	See *p*-tolylacetaldehyde			
α-Methylphenylacetaldehyde	See 2-phenylpropionaldehyde			
Methyl phenylacetate	methyl *α*-toluate	2733	FDA §172.515	Synthetic flavor
2-Methylphenyl acetate	See *o*-tolyl acetate			
p-Methylphenylbenzene	See 4-methylbiphenyl			
2-Methyl-4-phenylbutanal	See 2-methyl-4-phenylbutyraldehyde			
3-Methyl-4-phenyl-3-butene-2-one	benzylidene acetone methyl	2734	FDA §172.515	Synthetic flavor
2-Methyl-4-phenyl-2-butyl acetate	dimethylphenylethyl carbinyl acetate	2735	FDA §172.515	Synthetic flavor
2-Methyl-4-phenyl-2-butyl isobutyrate	dimethylphenylethyl carbinyl isobutyrate	2736	FDA §172.515	Synthetic flavor
2-Methyl-4-phenyl-2-butyl 2-methylpropanoate	See 2-methyl-4-phenyl-2-butyl isobutyrate			

2-Methyl-4-phenylbutyraldehyde		2737		
3-Methyl-2-phenylbutyraldehyde	α-isopropyl phenylacetaldehyde	2738	FDA §172.515	Synthetic flavor
Methyl 4-phenylbutyrate		2739	FDA §172.515	Synthetic flavor
Methyl γ-phenylbutyrate	See methyl 4-phenylbutyrate			
Methyl phenyl carbinol	See α-methylbenzyl alcohol			
Methyl phenylcarbinyl acetate	See α-methylbenzyl acetate			
Methyl phenylcarbinyl butyrate	See α-methylbenzyl butyrate			
Methyl phenylcarbinyl formate	See α-methylbenzyl formate			
Methyl phenylcarbinyl isobutyrate	See α-methylbenzyl isobutyrate			
Methyl phenylcarbinyl propionate	See α-methylbenzyl propionate			
Methyl phenyl diketone	See 1-phenyl-1,2-propanedione			
4-Methyl-2-phenyl-*m*-dioxolane	See benzaldehyde propylene glycol acetal			
4-Methyl-2-phenyl-1,3-dioxolane	See benzaldehyde propylene glycol acetal			
p-Methylphenyl dodecanoate	See *p*-tolyl laurate			
Methyl phenyl ether	See anisole			
Methyl 2-phenylethyl carbinol	See 4-phenyl-2-butanol			
2-Methyl-2-phenylethyl 2-methylpropanoate	See 2-phenylpropyl isobutyrate			
3-Methyl-3-phenylglycidic acid ethyl ester	Aldehyde C-16; ethyl methylphenylglycidate; so-called strawberry aldehyde		FDA §182.1	GRAS, synthetic flavor and adjuvant
Methyl phenyl glyoxal	See 1-phenyl-1,2-propanedione			
5-Methyl-2-phenyl-2-hexenal		3199		
p-Methylphenyl isobutyrate	See *p*-tolyl isobutyrate			
Methyl phenyl ketone	See acetophenone			

REGULATORY STATUS OF DIRECT FOOD ADDITIVES (continued)

Material	Alternatate Name [Source]	FEMA No.	Regulation	Limitations
p-Methylphenyl 3-methylbutyrate	See *p*-tolyl 3-methylbutrate			
Methyl phenylmethyl disulfide	See methyl benzyldisulfide			
p-Methylphenyl 2-methylpropanoate	See *p*-tolyl isobutyrate			
2-Methylphenyl 2-methylpropanoate	See *o*-tolyl isobutyrate			
p-Methylphenyl octanoate	See *p*-tolyl octanoate			
3-Methyl-1-phenyl-3-pentanol	See 1-phenyl-3-methyl-3-pentanol			
4-Methyl-1-phenyl-2-pentanol	See α-isobutylphenethyl alcohol			
4-Methyl-1-phenyl-2-pentanone	Benzyl isobutyl ketone	2740	FDA §172.515	Synthetic flavor/adjuvant
4-Methyl-2-phenyl-2-pentenal		3200		
p-Methylphenyl phenylacetate	See *p*-folyl phenylacetate			
2-Methyl-1-phenyl-2-propanol	See α,α-dimethylphenethyl alcohol			
2-(4-Methylphenyl)-2-propanol	See *p*,α,α-trimethylbenzyl alcohol			
2-Methyl-3-phenyl-2-propenal	See α-methylcinnamaldehyde			
3-(*p*-Methylphenyl) propenal	See *p*-methylcinnamaldehyde			
Methyl 3-phenylpropenoate	See methyl cinnamate			
2-(4-Methylphenyl) propionaldehyde	See 2-(*p*-tolyl)-propionaldehyde			
Methyl 3-phenylpropionate	Methyl hydrocinnamate	2741	FDA §172.515	Synthetic flavor
p-Methylphenyl α-toluate	See *p*-tolyl phenylacetate			

Methyl polysiloxane			MID	To retard foaming in curing pickles, at 50 ppm
2-Methyl propanal	See isobutyraldehyde			
Methyl propanoate	See methyl propionate			
2-Methylpropanoic acid	See isobutyric acid			
2-Methyl-1-propanol	See isobutyl alcohol			
2-Methyl-2-propen-1-yl butyrate	See 2-methylallyl butyrate			
Methyl 1-propenyl disulfide	1-Propenyl methyl disulfide	3576		See *Food Technol.*, February 1978
	Methyldithio-1 propene	3576		See *Food Technol.*, February 1978
Methyl propenyl disulfide		—		
2-Methyl-2-propenyl hexanoate	See methyallyl caproate			
Methyl propionate		2742	FDA §172.515	Synthetic flavor
2-Methyl-1-propyl acetate	See isobutyl acetate			
Methylpropylacetic acid	See 2-methylvaleric acid			
2-Methyl-1-propyl acetoacetate	See isobutyl acetoacetate			
Methyl β-propylacrylate	See methyl 2-hexenoate			
2-Methylpropyl benzoate	See isobutyl benzoate			
2-Methylpropyl benzyl carbinol	See α-isobutylphenethyl alcohol			
2-Methyl-1-propyl butyrate	See isobutyl butyrate			
2-Methyl-1-propyl caproate	See isobutyl hexanoate			
Methyl *n*-propyl carbinol	See 2-pentanol			
2-Methylpropyl cinnamate	See isobutyl cinnamate			
1-Methyl-4-propyl(iso)-1,3-cyclohexadiene	See α-terpinene			
1-Methyl-4-propyl(iso)-1,4-cyclohexadiene	See γ-terpinene			
1-Methyl-4-propyl(iso)-1,5-cyclohexadiene	See α-phellandrene			
2-(1-Methylpropyl)-cyclohexanone	See 2-*sec*-butylcyclohexanone			
1-Methyl-4-propyl(iso)-1-cyclohexen-8-ol	See α-terpineol			
1-Methyl-5-*n*-propyl-1-cyclohexen-3-one	See 3-methyl-5-propyl-2-cyclohexen-1-one			

REGULATORY STATUS OF DIRECT FOOD ADDITIVES (continued)

Material	Alternatate Name [Source]	FEMA No.	Regulation	Limitations
3-Methyl-5-propyl-2-cyclohexen-1-one	celery ketone	3577	FDA §172.515	Synthetic flavor/adjuvant; see *Food Technol*, February 1978
Methyl propyl diketone	See 2,3-hexanedione			
Methyl propyl(iso) diketone	See 4-Methyl-2,3-pentanedione			
Methyl propyl disulfide		3201		
2-Methyl-1-propyl formate	See isobutyl formate			
2-Methyl-1-propyl heptanoate	See isobutyl heptanoate			
2-Methylpropyl *o*-hydroxybenzoate	See isobutyl salicylate			
Methyl propyl ketone	See 2-pentanone			
2-Methylpropyl 2-*N*-methylaminobenzoate	See isobutyl *N*-methylanthranilate			
2-Methylpropyl 3-methylbutyrate		3369		
2-Methyl-1-propyl 2-methylpropanoate	See isobutyl isobutyrate			
2-Methylpropyl phenylacetate	See isobutyl phenylacetate			
2-Methylpropyl *β*-phenylacrylate	See isobutyl cinnamate			
Methyl *p*-propylphenyl ether	See *p*-propyl anisole			
2-Methylpropyl 3-phenylpropenoate	See isobutyl cinnamate			
2-Methyl-1-propyl propanoate	See isobutyl propionate			
2-(2-Methylpropyl) pyridine		3370		
3-(2-Methylpropyl) pyridine		3371		
2-Methyl-1-propyl salicylate	See isobutyl salicylate			
2-(1-Methylpropyl) thiazole		3372		

Methyl propyl trisulfide	propyl methyl trisulfide	3308		See *Food Technol.*, May 1972
Methylprotocatechuic aldehyde	See vanillin			
2-Methylpyrazine	2-methyl-1,4-diazine	3309		See *Food Technol.*, May 1972
Methyl pyrazinyl ketone	See acetylpyrazine			
Methyl pyridyl ketone	See 3-acetylpyridine			
Methyl 2-pyridyl ketone	See 2-acetylpyridine			
Methyl 3-pyridyl ketone	See 3-acetylpyridine			
Methyl *β*-pyridyl ketone	See 3-acetylpyridine			
Methyl pyromucate	See methyl 2-furoate			
2-Methyl-4-pyron-3-yl 2-methylpropanoate	See maltyl isobutyrate			
Methyl 2-pyrrolyl ketone	2-acetylpyrrole	3202		
p-Methylquinoline	See 6-methylquinoline			
4-Methylquinoline	See lepidine			
6-Methylquinoline		2744	FDA §172.515	Synthetic flavor
5-Methylquinoxaline		3203		
Methyl salicylate	wintergreen oil	2745		
Methyl salicylate *O*-methyl ether	See methyl *o*-methoxybenzoate			
Methyl styryl carbinol	See 4-phenyl-3-buten-2-ol			
Methyl styryl ketone	See 4-phenyl-3-buten-2-one			
Methyl succinate	See dimethyl succinate			
Methyl sulfhydrate	See methyl mercaptan			
Methyl sulfide	dimethyl sulfide	2746	FDA §172.515	Synthetic flavor
Methyl tetradecanoate	See methyl myrislate			
2-Methyltetrahydrofuran-3-one		3373		
2-Methyltetrahydrothiophen-3-one		3512		
Methyltheobromine	See caffeine			
4-Methyl-5-thiazoleethanol	4-methyl-5-(*β*-hydroxyethyl)-thiazole	3204		
4-Methyl-5-thiazoleethanol acetate		3205		
2-Methylthioacetaldehyde		3206		
3-(Methylthio)butanal		3374		
4-(Methylthio)butanal		3414		
1-(Methylthio)-2-butanone		3207		

REGULATORY STATUS OF DIRECT FOOD ADDITIVES (continued)

Material	Alternatate Name [Source]	FEMA No.	Regulation	Limitations
4-(Methylthio)-2-butanone		3375		
3-(Methylthio)butyraldehyde	See 3-(methylthio)butanal			
4-(Methylthio)butyraldehyde	See 4-(methylthio)butanal			
Methyl thiobutyrate	methanethiol *n*-butyrate; thiobutyric acid, methyl ester	3310		See *Food Technol.*, May 1972
Methyl thiofuroate	thiofuroic acid, methyl ester	3311		See *Food Technol.*, May 1972
3-(Methylthio)-1-hexanal		3438		
2-Methylthiolan-3-one	See 2-methyltetrahydrothiophen-3-one			
2-Methyl-5-thiomethylfuran	See 2-methyl-5-(methylthio)furan			
4-(Methylthio)-4-methyl-2-pentanone		3376		
(Methylthio)methylpyrazine (mixture of isomers)	2-methyl-3-, 5-, or 6-methylthiopyrazine	3208		10 ppm max in finished food
5-(Methylthio)-2-(methylthio)methylpent-2-en-1-al	See 2,8-dithianon-4-en-4-carboxaldehyde			
5-Methyl-2-thiophenecarboxaldehyde		3209		
o-(Methylthio)phenol	2-(methylthio)phenol; thioguaiacol	3210		
2-Methylthiophenol	See *o*-toluenethiol	·		
2-(Methylthio)phenol	See *o*-(methylthio)phenol			
3-(Methylthio)propanol		3415		
3-Methylthiopropionaldehyde	methional; β-methiopropionaldehyde; 2-methylthiopropionaldehyde	2747	FDA §172.515	Synthetic flavor
β-(Methylthio)propionaldehyde	See 3-(methylthio)propionaldehyde			
3-(Methylthio)propyl alcohol	See 3-(methylthio)propanol			
3-Methylthiopropyl isothiocyanate		3312		See *Food Technol.*, May 1972

α-Methyltolualdehyde	See 2-phenylpropionaldehyde			
Methyl α-toluate	See methyl phenylacetate			
Methyl *p*-tolyl carbinol	See *p*,α-dimethylbenzyl alcohol			
Methyl *o*-tolyl ether	See *o*-methylanisole			
Methyl *p*-tolyl ether	See *p*-methylanisole			
Methyl *p*-tolyl ketone	See 4′-methylacetophenone			
2-Methyl-3-tolyl propionaldehyde (mixed *o*-, *m*-, *p*-)		2748	FDA §172.515	Synthetic flavor
Methyl trisulfide	See dimethyl trisulfide			
2-Methylundecanal	aldehyde C-12, M.N.A.; methyl nonyl acetaldehyde	2749	FDA §172.515	Synthetic flavor
Methyl 9-undecenoate		2750	FDA §172.55	Synthetic flavor
Methyl undecylenate	See methyl 9-undecenoate			
Methyl undecyl ketone	See 2-tridecanone			
Methyl 2-undecynoate	methyl decyne carbonate	2751	FDA §172.515	Synthetic flavor
2-Methylvaleraldehyde	See 2-methylpentanal			
Methyl valerate		2752	FDA §172.515	Synthetic flavor
Methyl valerianate	See methyl valerate			
2-Methylvaleric acid	2-methylpentanoic acid	2754	FDA §172.515	Synthetic flavor
3-Methylvaleric acid	See 3-methylpentanoic acid			
4-Methylvaleric acid	See 4-methylpentanoic acid			
β-Methylvaleric acid	See 3-methylpentanoic acid			
5-Methyl-δ-valerolactone	See δ-hexalactone			
Methyl vanillin	See veratraldehyde			
trans-8-Methyl-*N*-vanillyl-6-nonenamide	See *N*-(4-hydroxy-3-methoxybenzyl)-8-methyl-6-nonenamide			
Methylvinyl homogeranyl carbinol	See nerolidol			
2-Methyl-5-vinylpyrazine		3211		10 ppm max in finished food
2-(1-Methylvinyl)pyrazine	See isopropenylpyrazine			
2-(α-Methylvinyl)pyrazine	See isopropenylpyrazine			
4-Methyl-5-vinylthiazole		3313		See *Food Technol.*, May 1972
Micromet®	See sodium hexametaphosphate			
Microwave radiation for heat treatment of food			FDA §121.3008	Treatment of whole-fish protein concentrate complying with §121.1202 for reduction of solvent residues; microwave frequency authorized by FCC regulations; microwave power, 25 W max

REGULATORY STATUS OF DIRECT FOOD ADDITIVES (continued)

Material	Alternate Name [Source]	FEMA No.	Regulation	Limitations
Milfoil	See yarrow, herb [*Achillea millefolium* (L.)]			
Milk protein hydrolysate			MID	To flavor various meat products
Mimosa, flower, absolute	Black wattle [*Acacia decurrens* Willd. var. *dealbata*]	2755	FDA §172.510	Natural flavor
Mineral oil			FDA §121.1099	In defoaming agents limited to use in processing beet sugar and yeast; up to 150 ppm in yeast, measured as hydrocarbons
MMS	See *dl*-(3-amino-3-carboxypropyl)dimethylsulfonium chloride			
M.N.K.	See 2-undecanone			
Modified hops extract	See hops extract, modified			
Molasses			FDA §8.303	In production of caramel
			FDA §8.301	In dried algae meal used in chicken feed
Molasses extract	[*Saccharum officinarum* (L.)]		FDA §182.1	GRAS, natural flavor extractive
Monoammonium glutamate			FDA §182.1	GRAS, miscellaneous and/or general-purpose food additive
Mono- and diglycerides	See glycerides (mono- and di-)			
Monocalcium phosphate			FDA §29.4 and §29.5	In artificially sweetened fruit jellies and preserves
			FDA §51.990	In specified canned vegetables
			FDA §17	In bakery products, up to 0.75% by weight of flour
			FDA §15.50 and §15.60	Acid-reacting agent in self-rising flour and enriched self-rising flour
			FDA §15.70	0.25 to 0.75% in phosphated flour
			FDA §15.506 and §15.507	Acid-reacting agent in self-rising, white corn meal and self-rising, yellow corn meal in combination with sodium bicarbonate, so that combination is not more than 4.5% of corn meal

Monoglyceride citrate			FDA §121.1036	Synergist and solubilizer in antioxidant formulations for addition to oils and fats, up to 200 ppm of combined weight of oil or fat and the additive; product specifications apply
			MID	To increase effectiveness of antioxidants in lard and shortening, at 0.01%
Monoisopropyl citrate			FDA §182.1	GRAS, sequestrant
			FDA	0.02% max in margarine by weight of finished margarine
			MID	To increase effectiveness of antioxidants in lard, shortening, and oleomargarine; 0.01% in lard and shortening; 0.02% in oleomargarine
Monoolein	See glyceryl monooleate			
Monopotassium glutamate			FDA §182.1	GRAS, miscellaneous and/or general-purpose food additive
Monosodium glutamate		2756	FDA §121.23	GRAS
			MID	To flavor various meat products
Monosodium orthophosphate	Monosodium *o*-phosphate		FDA §121.1031	Esterification of food starch; residual phosphate in food, 0.4% max calculated as phosphorus
Monosodium phosphate			FDA §19	Emulsifier, up to 3% by weight of specified cheeses
			MID	To decrease amount of cooked-out juices in cured hams, pork shoulder picnics, and loins, canned hams and pork shoulder picnics, chopped ham, and bacon; 5.0% of phosphate in pickle at 10% pump level; 0.5% of phosphate in product; only clear solution may be injected into product
Monosodium phosphate derivatives			FDA §182.1	Monosodium phosophate derivatives of mono- and di-glycerides of edible fats or oils or edible fat-forming fatty acids; GRAS, emulsifiers
Monostearin	See glyceryl monostearate			
α-Monostearin	See glyceryl monostearate			

REGULATORY STATUS OF DIRECT FOOD ADDITIVES (continued)

Material	Alternate Name [Source]	FEMA No.	Regulation	Limitations
Morellone	See 3-benzyl-4-heptanone			
Mountain maple, bark	[*Acer Spicatum* Lam.]	—	FDA §172.510	
Mountain maple, extract solid	[*A. Spicatum* Lam.]	2757	FDA §172.510	
Mucara	See karaya, gum [*Sterculia urens* Roxb.]			
Mullein, flower	[*Verbascum phlomoidea* (L.); *V. thapsiforme* Schrad.]	—	FDA §172.510	Natural flavor in alcoholic beverages only
Muscolactone	See ω-pentadecalactone			
d,l-Muscone	See 3-methyl-1-cyclopentadecanone			
Musk	Tonquin musk [*Moschus moschiferus* (L.), musk deer]		FDA §182.1	GRAS, natural extractive
Musk ambrette	2,6-dinitro-3-methoxy-1-methyl-4-*tert*-butylbenzene	2758		
Musk tonquin	[*M. moschiferus* (L.)]	2759	FDA §182.1	Natural flavor
Musk yarrow	See iva, extract [*Achillea moschata* Jacq.]			
Mustard, black or brown	[*Brassica nigra* (L.) Koch; *B. juncea* (L.) Coss.]	2760	FDA §182.1	GRAS, natural flavor
Mustard oil	See allyl isothiocyanate			
Mustard, white or yellow	[*Brassica hirta* Moench.; *B. alba* (L.) Boiss]	2761	FDA §182.1	GRAS, natural flavor
Myrcene	7-methyl-3-methylene-1,6-octadiene	2762	FDA §172.515	Synthetic flavor
Myrcia oil	See bay leaf, West Indian, oil [*Pimenta racemosa* (Mill.) J. W. Moore; *P. acris* Kostel]			
Myristaldehyde	Aldehyde C-14 myristic; tetradecanal	2763	FDA §172.515	Synthetic flavor
Myristic acid	Tetradecanoic acid	2764	FDA §121.1070	
Myrrh, gum	[*Commiphora molmol* Engl.; *C. abyssinica* (Berg) Engl.; other *Commiphora* spp.]	2765	FDA §172.510	Natural flavor
Myrrh, oil	[*C. molmol* Engl.; *C. abyssinica*	2766	FDA §172.510	Natural flavor

	(Berg) Engl.; other *Commiphora* spp.]			
Myrtenal	See 2-formyl-6,6-dimethylbicyclo (3.1.1) hept-2-ene			
Myrtenol		3439		
Myrtenyl formate	See 2-hydroxymethyl-6,6-dimethylbicyclo-(3.1.1)hept-2-enyl formate			
Myrtle leaves	[*Myrtus communis* (L.)]	—	FDA §172.510	Natural flavor in alcoholic beverages only
Naphtha			FDA §8.390	Diluent in mixtures for coloring egg shells; no penetration of mixture into egg
2-Naphthalenthiol	2-Mercaptonaphthalene; 2-naphthyl mercaptan; 2-thionaphthol	3314		See *Food Technol.*, May 1972
2-Naphthyl anthranilate	See β-naphthyl anthranilate			
β-Naphthyl anthranilate	2-Naphthyl anthranilate	2767		
β-Naphthyl ethyl ether	2 -Ethoxynaphthalene	2768		
β-Naphthyl isobutyl ether		—		
2-Naphthyl mercaptan	See 2-*naphthalenthiol*			
β-Naphthyl methyl ether	Nerolin I; yara yara			
β-Naphthyl methyl ketone	See methyl β-naphthyl ketone			
Narcissin	See *p*-tolyl phenylacetate			
Naringin	[*Citrus paradisi* Macf.]	2769	FDA §182.1	GRAS, natural flavor extractive
Natural rubber (smoked sheet and latex solids)	[*Hevea brasiliensis*]		FDA §121.1059	Natural (coagulated or concentrated latex) masticatory substance in chewing-gum base
Nauli "gum"	See anethole			
Neofolione	See methyl 2-nonenoate			
d-Neomenthol	2-Isobutyl-5-methyl cyclohexanol; *d*-neo-menthol	2666	FDA §172.515	Synthetic flavor
Neral	See citral			
Neraniol	See nerol			
Nergenol	See nerol			
Nerodol	See nerol			
Nerol	*cis*-3,7-Dimethyl-2,6-octadien-1-ol	2770	FDA §172.515	Synthetic flavor
Neroli bigarade, oil	[*Citrus aurantium* (L.) subsp. *amara* (L.)]	2771	FDA §182.1	GRAS, natural flavor extractive

REGULATORY STATUS OF DIRECT FOOD ADDITIVES (continued)

Material	Alternate Name [Source]	FEMA No.	Regulation	Limitations
Nerolidol	3,7,11-Trimethyl-1,6,10-dodecatrien-3-ol	2772	FDA §172.515	Synthetic flavor
Nerolin I	See β-naphthyl methyl ether			
Nerolol	See nerol			
Nerosol	See nerol			
Neryl acetate		2773	FDA §172.515	Synthetic flavor
Neryl butyrate		2774	FDA §172.515	Synthetic flavor
Neryl ethanoate	See neryl acetate			
Neryl formate		2776	FDA §172.515	Synthetic flavor
Neryl isobutyrate		2775	FDA §172.515	Synthetic flavor
Neryl isovalerate		2778	FDA §172.515	Synthetic flavor
Neryl isovalerianate	See neryl isovalerate			
Neryl methanoate	See neryl formate			
Neryl 3-methylbutanoate	See neryl isovalerate			
Neryl 2-methylpropanoate	See neryl isobutyrate			
Neryl propanoate	See neryl propionate			
Neryl propionate		2777	FDA §172.515	Synthetic flavor
Nevitin	See *d*-sorbitol			
Niacin			FDA §182.1	GRAS, nutrient and/or dietary supplement
Niacinamide			FDA §182.1	GRAS, nutrient and/or dietary supplement
Nicotinamide ascorbic acid complex			FDA §121.1095	Source of ascorbic acid and nicotinamide in multivitamin preparations; product specifications apply
Niger gutta	[*Fucus platyphylla* Del.]		FDA §121.1059	Natural (coagulated or concentrated latex) masticatory substance in chewing-gum base
Nispero	[*Manilkara zapotilla* Gilly; *M. chicle* Gilly]		FDA §121.1059	Natural (coagulated or concentrated latex) masticatory substance in chewing-gum base
Nitrogen			FDA §182.1	GRAS, miscellaneous and/or general-

				purpose food additive
			MID	To exclude oxygen from sealed containers
Nitrosyl chloride			FDA §15	Bleaching substance for flour
Nitrous oxide	Dinitrogen monoxide	2779	FDA §182.1	GRAS, propellant for certain dairy and vegetable-fat toppings in pressurized containers
			FDA §121.1065	With octafluorocyclobutane as propellant and aerating agent for foamed or sprayed food products, except those standardized foods that do not provide for such use
			FDA §121.1180	With chloropentafluoroethane as propellant and aerating agent for foamed or sprayed foods
No-Doz®	See caffeine			
2,4-Nonadienal		3212		
Nona-2-*trans*, 6-*cis*-dienal		3377		
2,6-Nonadienal diethyl acetal		3378		
2,6-Nonadien-1-ol		2780	FDA §172.515	Synthetic flavor
γ-Nonalactone	Aldehyde C-18 (so-called); 4-hydroxynonanoic acid, γ-lactone	2781	FDA §172.515	Synthetic flavor
δ-Nonalactone	See hydroxynonanoic acid, δ-lactone			
Nonalol	See nonyl alcohol			
Nonamethylene dimercaptan	See 1,9-nonanedithiol			
n-Nonanal	Pelargonic aldehyde; aldehyde C-18	2782	FDA §172.515	Synthetic flavor
Nonane diacetate	See 1,3-nonanediol acetate (mixed esters)			
Nonanediol-1,4-acetate		—		
1,3-Nonanediol acetate (mixed esters)	Octyl crotonyl acetate	2783	FDA §172.515	Synthetic flavor
1,9-Nonanedithiol		3513		
Nonanoic acid	Pelargonic acid	2784	FDA §172.515	Synthetic flavor
Nonanoic aldehyde	See nonanal			
1-Nonanol	See nonyl alcohol			
2-Nonanol	Methyl *n*-heptyl carbinol	3315		See *Food Technol.*, May 1972

REGULATORY STATUS OF DIRECT FOOD ADDITIVES (continued)

Material	Alternate Name [Source]	FEMA No.	Regulation	Limitations
Nonanolide-1,4	See γ-nonalactone			
Nonan-2-one	See 2-nonanone			
2-Nonanone	Methyl heptyl ketone	2785	FDA §172.515	Synthetic flavor
3-Nonanone		3440		
3-Nonanon-1-yl acetate	1-Hydroxy-3-nonanone acetate	2786	FDA §172.515	Synthetic flavor
Nonanoyl 4-hydroxy-3-methoxybenzylamide	See *N*-(4-hydroxy-3-methoxybenzyl) nonanamide			
N-Nonanoyl vanillylamide	See nonanoyl 4-hydroxy-3-methoxy-benzylamide			
2-Nonenal		3213		
6-Nonenal		—		
cis-6-Nonenol	See *cis*-6-nonen-1-ol			
cis-6-Nonen-1-ol		3465		
trans-2-Nonen-1-ol		3379		
Nonoic acid	See nonanoic acid			
Nonoic aldehyde	See nonanal			
Nonyl carbinol	See 1-decanal			
Nonyl acetate	Acetate C-9	2788	FDA §172.515	Synthetic flavor
3-Nonylacrolein	See 2-dodecanal			
Nonyl alcohol	2,6-Dimethyl heptanol-4; 1-nonanol	2789	FDA §172.515	Synthetic flavor
Nonyl caprylate	See nonyl octanoate			
Nonyl ethanoate	See nonyl acetate			
Nonylic acid	See nonanoic acid			
Nonyl isopentanoate	See nonyl isovalerate			
Nonyl isovalerate		2791	FDA §172.515	Synthetic flavor
Nonyl isovalerianate	See nonyl isovalerate			
γ-Nonyllactone	See γ-nonalactone			
Nonyl octanoate		2790	FDA §172.515	Synthetic flavor
Nonyl octylate	See nonyl octanoate			
Nootkatone	5,6-Dimethyl-8-	3166		

	isopropenylbicyclo(4.4.0)-dec-1-en-3-one; 4,4a,5,6,7,8-hexahydro-6-isopropenyl-4,4a-dimethyl-2(3*H*)-naphthalenone			
Nopinene	See β-pinene			
Nutmeg	[*Myristica fragrans* Houtt.]	2792	FDA §182.1	GRAS, natural flavor and natural flavor extractive
Nutmeg, oil	[*M. fragrans* Houtt.]	2793	FDA §182.1	Natural flavor
Oak, chips, white	[*Quercus alba* (L.)]	2794	FDA §172.510	Natural flavor
Oak moss, absolute	[*Evernia prunastri* (L.) Ach.; *E. furfuracea* (L.) Mann; and other lichens]	2795	FDA §172.510	Natural flavor; thujone-free
Oak, wood, English	[*Q. robur* (L.)]		FDA §172.510	Natural flavor in alcoholic beverages only
Oat gum			FDA §19 FDA-GRAS	0.8% max in specified cheese products
Ocimene	3,7-Dimethyl-1,3,6-octatriene; *trans*-β-ocimene	—	FDA §172.515	Synthetic flavor
trans-β-Ocimene	See ocimene			
Octaacetyl sucrose	See sucrose octaacetate			
9,12-Octadecadienoic acid (48%) and 9,12,15-octadecatrienoic acid (52%)		3380		
Octadecanoic acid	See stearic acid			
9-Octadecenoic acid	See oleic acid			
Octadecylic acid	See stearic acid			
2-*trans*-6-*trans*-Octadienal		3466		
3,6-Octadien-1-ol			FDA §182.1	GRAS, synthetic flavor
Octafluorocyclobutane	Perfluorocyclobutane		FDA §121.1065	Alone or with carbon dioxide and/or nitrous oxide and/or propane, as propellant and aerating agent for foamed or sprayed food products, except standardized foods that do not provide for such use; product specifications apply
			FDA §121.1181	With chloropentafluoroethane as propellant and aerating agent for foamed or sprayed food products; complying with §121.1065

REGULATORY STATUS OF DIRECT FOOD ADDITIVES (continued)

Material	Alternate Name [Source]	FEMA No.	Regulation	Limitations
1,2,3,5,6,7,8,8a-Octahydro-1,8a-dimethyl-7-(1-methylethenyl) naphthalene	See valencene			
Octalactone			FDA §172.515	Synthetic flavor
γ-Octalactone	4-Hydroxyoctanoic acid, γ-lactone	2796	FDA §172.515	Synthetic flavor
δ-Octalactone		3214		
Octaldehyde dimethyl acetal	See octanal dimethyl acetal			
Octamethylene dimercaptan	See 1,8-octanedithiol			
n-Octanal	Aldehyde C-8; caprylaldehyde	2797	FDA §172.515	Synthetic flavor
Octanal dimethyl acetal	Aldehyde C-8 dimethyl acetal	2798	FDA §172.515	Synthetic flavor
1,8-Octanedithiol		3514		
Octanoic acid	Caprylic acid	2799	FDA §172.515	Synthetic flavor
1-Octanol	See 1-octyl alcohol			
2-Octanol	See 2-octyl alcohol			
3-Octanol	See 3-octyl alcohol			
Octanolide-1,4	See γ-octalactone			
5-Octanol-4-one	See 5-hydroxy-4-octanone			
2-Octanone	Methyl hexyl ketone	2802	FDA §172.515	Synthetic flavor
3-Octanone	Ethyl amyl ketone	2803	FDA §172.515	Synthetic flavor
3-Octanon-1-ol	Methylol methyl amyl ketone	2804	FDA §172.515	Synthetic flavor
2-Octenal		3215		
6-Octenal		—		
3-Octenol	See 1-octen-3-ol			
cis-3-Octenol	See *cis*-3-octen-1-ol			
1-Octen-3-ol	Amyl vinyl carbinol	2805	FDA §172.515	Synthetic flavor
cis-3-Octen-1-ol		3467		
1-Octen-3-one		3515		
3-Octen-2-one		3416		
Octenyl acetate	See 1-octen-3-yl acetate			

β-Octenyl acetate	See 1-octen-3-yl acetate			
1-Octen-3-yl acetate	3-Acetoxyoctene; amyl vinyl carbinyl acetate	3582	FDA §172.515	Synthetic flavor/adjuvant; see *Food Technol.*, February 1978
trans-2-Octen-1-yl acetate		3516		
trans-2-Octen-1-yl butanoate		3517		
trans-2-Octen-1-yl butyrate	See *trans*-2-octen-1-yl butanoate			
1-Octenyl succinic anhydride			FDA §121.1031	Modifier of food starch 3.0% max; up to 2.0% in combination with 2.0% max of aluminum sulfate
Octoic acid	See octanoic acid			
1-Octen-3-yl acetate	3-Acetoxyoctene	3582		See *Food Technol.*, February 1978
Octyl acetate	2-Ethyl hexyl acetate	2806	FDA §172.515	Synthetic flavor
3-Octyl acetate	Amyl ethyl carbinyl acetate	3583	FDA §172.515	Synthetic flavor/adjuvant; see *Food Technol.*, February 1978
Octyl alcohol	See 1-octyl alcohol			
1-Octyl alcohol	1-Octanol	2800	FDA §172.515	Synthetic flavor
2-Octyl alcohol	2-Octanol	2801	FDA §172.515	Synthetic flavor
3-Octyl alcohol	3-Octanol	—	FDA §172.515	Synthetic flavor/adjuvant
pri-Octyl alcohol	See 1-octyl alcohol			
sec-Octyl alcohol	See 2-octyl alcohol			
Octyl alcohol, secondary	See 2-octyl alcohol			
Octylaldehyde	See octanal			
Octyl butanoate	See octyl butyrate			
Octyl butyrate		2807	FDA §172.515	Synthetic flavor
γ-Octyl-γ-butyrolactone	See γ-dodecalactone			
γ-*n*-Octyl-γ-butyrolactone	See γ-dodecalactone			
Octyl caprylate	See octyl octanoate			
Octyl carbinol	See nonyl alcohol			
Octyl crotonyl acetate	See 1,3-nonanediol acetate (mixed esters)			
Octyl ethanoate	See octyl acetate			
Octyl formate		2809	FDA §172.515	Synthetic flavor
Octyl2-furancarboxylate	See octyl 2-furoate			
Octyl 2-furoate		3518		
Octyl heptanoate		2810	FDA §172.515	Synthetic flavor
Octyl heptoate	See octyl heptanoate			
Octyl heptylate	See octyl heptanoate			

REGULATORY STATUS OF DIRECT FOOD ADDITIVES (continued)

Material	Alternate Name [Source]	FEMA No.	Regulation	Limitations
4-*n*-Octyl-4-hydroxybutanoic acid lactone	See γ-dodecalactone			
Octylic acid	See octanoic acid			
Octyl isobutyrate		2808	FDA §172.515	Synthetic flavor
Octyl isopentanoate	See octyl isovalerate			
Octyl isovalerate		2814	FDA §172.515	Synthetic flavor
Octyl isovalerianate	See octyl isovalerate			
Octyl methanoate	See octyl formate			
Octyl 3-methylbutanoate	See octyl isovalerate			
Octyl 3-methylbutyrate	See octyl isovalerate			
Octyl 2-methylpropanoate	See octyl isobutyrate			
Octyl octanoate		2811	FDA §172.515	Synthetic flavor
Octyl octylate	See octyl octanoate			
Octyl phenylacetate		2812	FDA §172.515	Synthetic flavor
Octyl propanoate	See octyl propionate			
Octyl propionate		2813	FDA §172.515	Synthetic flavor
Octyl α-toluate	See octyl phenylacetate			
Oenanthal	See heptanal			
Oenanthal dimethyl acetal	See heptanal dimethyl acetal			
Oenanthic acid	See heptanoic acid			
Oenanthylic acid	See heptanoic acid			
Oenanthylidene acetone	See 3-decen-2-one			
Oil of Niobe	See methyl benzoate			
Oleic acid	*cis*-9-Octadecenoic acid	2815	FDA §121.1070	
(mono) Olein	See glyceryl monooleate			
Oleinic acid	See oleic acid			
Olibanum, oil	Frankincense [*Boswellia carterii* Birdw. and other *Boswellia* spp.]	2816	FDA §172.515	Synthetic flavor
Olive, oil	[*Olea europaea* (L.)]		FDA-GRAS	
Onion, oil	[*Allium cepa* (L.)]	2817	FDA §182.1	GRAS, natural flavor extractive
Opopanax, gum and oil	Bisabol myrrh [*Opopanax chironium* Koch. (true opopanax) or *Commi-*	—	FDA §172.510	Natural flavor

	phora erythraea Engl. var. *glabrescens*]			
Orange B			FDA §8.202	Coloring casings or surfaces of frankfurters and sausages, 150 ppm max; product specifications apply
Orange, bitter, flower and peel	See Curaçao peel, extract [*Citrus aurantium* (L.)]			
Orange blossoms, absolute	[*C. aurantium* (L.)]	2818	FDA §182.1	GRAS, natural flavor extractive
Orange flowers	[*C. aurantium* (L.)]	2819		
Orange flowers, bitter, oil	See neroli bigarde, oil [*C. aurantium* (L.)]			
Orange leaf, absolute	[*C. aurantium* (L.)]	2820	FDA §182.1	Natural flavor
Orange, oil, distilled	[*C. sinensis* (L.) Osbeck]	2821		
Orange, oil, terpeneless	[*C. sinensis* (L.) Osbeck]	2822		
Orange peel, bitter, oil	[*C. aurantium* (L.)]	2823	FDA §182.1	Natural flavor
Orange, peel, sweet, extract	[*C. sinensis* (L.) Osbeck; *C. aurantium* var. *dulcis* (L.)]	2824	FDA §182.1	GRAS, natural flavor extractive
Orange peel, sweet, oil	[*C. sinensis* (L.) Osbeck; *C. aurantium* var. *dulcis* (L.)]	2825	FDA §182.1	Natural flavor
Orange peel, sweet, oil, terpeneless	[*C. sinensis* (L.) Osbeck; *C. aurantium* var. *dulcis* (L.)]	2826	FDA §182.1	Natural flavor
Oranger crystals	See methyl β-naphthyl ketone			
Oregano	Oreganum, Mexican oregano, Mexican sage, origan [*Lippia* spp.]	2827	FDA §182.1	GRAS, natural flavor
Oregano, Mexican	See oregano [*Lippia* spp.]			
Oreganum	See oregano			
Origan	See oregano			
Origanol	See 4-carvomenthenol			
Origanum, oil (extractive)	[*Origanum vulgare* (L.) and other *Origanum* spp.]		FDA §182.1	GRAS, natural flavor extractive
"Origanum," Spanish	See origanum, oil (extractive)			
Orris, concrete, liquid, oil	*Iris florentina* (L.)	2829		
Orris root, extract	[*I. germanica* (L.), including its variety *florentina* Dykes]	2830	FDA §172.510	Natural flavoring substances and natural substances used in conjunction with flavors
Ox bile extract	Oxgall		FDA §182.1	GRAS, emulsifying agent in dried egg whites, up to 0.1%

REGULATORY STATUS OF DIRECT FOOD ADDITIVES (continued)

Material	Alternate Name [Source]	FEMA No.	Regulation	Limitations
3-Oxobutanal, dimethyl acetal		3381		
α-Oxodiphenylmethane	See benzophenone			
3-Oxo-1-octanol	See 3-octanon-1-ol			
4-Oxopentanoic acid	See levulinic acid			
γ-Oxopentanoic acid	See levulinic acid			
1-Oxo-2-phenylethane	See phenylacetaldehyde			
2-Oxopropanal	See pyruvaldehyde			
2-Oxopropane	See acetone			
2-Oxopropanoic acid	See pyruvic acid			
Oxoundecane	See 2-undecanone			
4-Oxovaleric acid	See levulinic acid			
Oxydihydrocitronellal	See hydroxycitronellal			
Oxygen, active			FDA §121.1031	0.45% max, obtained from hydrogen peroxide and/or peracetic acid, as modifier for food starch; alone or in combination with chlorine as sodium hypochlorite (not to exceed 0.055 lb of chlorine per pound of dry starch) and not more than 25% of propylene oxide
1-Oxy-2-methoxybenzene	See guaiacol			
Oxysterin			FDA §121.1016	Crystallization inhibitor in cottonseed and soybean cooking and salad oils and in vegetable oils; as release agent in vegetable oils and vegetable shortening, up to 0.125% of combined weight of oil plus additive; product specifications apply
			FDA §121.1099	In defoaming agents limited to use in processing beet sugar and yeast
14-Oxytetradecane carbonic acid lactone	See ω-pentadecalactone			

P.A.D.M.A.	See phenylacetaldehyde dimethyl acetal			
Pale catechu	See catechu, powder [*Acacia catechu* Willd.]			
Palma Christi, oil	See castor, oil [*Ricinus communis* (L.)]			
Palmarosa, oil	East Indian geranium [*Cymbopogon martini* Stapf. var. *motia*]	2831	FDA §182.1	GRAS, natural flavor extractive
Palmitic acid	Hexadecanoic acid	2832	FDA §121.1070	
Palmityl alcohol	See 1-hexadecanol			
Pansy	[*Viola tricolor* (L.)]	—	FDA §172.510	Natural flavor in alcoholic beverages only
D-Pantothenamide			FDA §121.1123	Source of pantothenic acid activity in foods for special dietary use
D-Pantothenyl alcohol			FDA §182.1	GRAS, nutrient and/or dietary supplement
Papain	Papayotin		FDA §182.1	GRAS, miscellaneous and/or general-purpose food additive
			MID	To soften tissues of beef cuts, with water, salt, and monosodium glutamate; tissue weight gain, 3% max
Paprika	[*Capsicum annuum* (L.)]	2833	FDA §182.1	GRAS, natural flavor and natural flavor extractive
			FDA §8.307	Food coloring
Paprika, oleoresin	[*C. annuum* (L.)]	2834	FDA §8.308	Food coloring; solvent residues are limited; product specifications apply
			FDA §182.1	Natural flavor
Parabar	See butylated hydroxytoluene			
Paradol®	See methyl *p*-hydroxybenzoate			
Paraffin	See paraffin wax			
Paraffin wax	Paraffin	3216	FDA §121.1059	Synthetic masticatory substance in chewing-gum base; product specifications apply
Parsley	[*Petroselinum sativum* Hoffm.; *Carum petroselinum* Benth. and Hook; other *Petroselinum* spp.]	2835	FDA §182.1	GRAS, natural flavor and natural flavor extractive

REGULATORY STATUS OF DIRECT FOOD ADDITIVES (continued)

Material	Alternate Name [Source]	FEMA No.	Regulation	Limitations
Parsley, oil	[*P. sativum* Hoffm.; *C. petroselinum* Benth. and Hook; other *Petroselinum* spp.]	2836	FDA §182.1	Natural flavor
Parsley, oleoresin	[*P. sativum* Hoffm.; *C. petroselinum* Benth. and Hook; other *Petroselinum* spp.]	2837	FDA §182.1	Natural flavor
Passion flower	[*Passiflora incarnata* (L.)]	—	FDA §172.510	Natural flavor
Patchouly, oil	[*Pogostemon cablin* Benth.; *P. heyneanus* Benth.]	2838	FDA §172.510	Natural flavor
β-P.E.A.	See phenethyl alcohol			
Peach aldehyde	See γ-undecalactone			
Peach kernel	Persic oil, bitter almond [*Prunus persica* Sieb and Zucc.]		FDA §182.1	GRAS, solvent-free natural extractive used in conjunction with natural flavorings
Peach, leaf	[*P. persica* (L.) Batsch]	—	FDA §172.510	Natural flavor in alcoholic beverages only; not to exceed 25 ppm of prussic acid
Peanut stearine	[*Arachis hypogaea* (L.)]		FDA §182.1	GRAS, solvent-free natural extractive used in conjunction with natural flavoring
Pectin			FDA-GRAS	
Pectin sugar	See L-arabinose			
Pediococcus cerevisiae culture			MID	Starter to develop flavor in dry sausage, pork roll, thuringer, Lebanon bologna, cervelat, and salami, at 0.5%
Pegu catechu, extract	See catechu, extract [*Acacia catechu* Willd.]			
γ-Pelargolactone	See γ-nonalactone			
Pelargonaldehyde	See *n*-nonanal			
Pelargonic acid	See nonanoic acid			
Pelargonic alcohol	See nonyl alcohol			
Pelargonic aldehyde	See *n*-nonanal			

Pelargonyl vanillylamide	See *N*-(4-hydroxy-3-methoxybenzyl)nonanamide			
P.E.M.E.C.	See 1-phenyl-3-methyl-3-pentanol			
Pendare	[*Couma macrocarpa* Barb. Rodr.; *C. utilis* (Mart.) Muell. Arg.]		FDA §121.1059	Natural (coagulated or concentrated latex) masticatory substance in chewing-gum base
Penicillium roqueforti			FDA §19	Mold in specified cheeses
Pennyroyal, American and European, oil	[*Hedeoma pulegioides* (L.) Pers.; *Mentha pulegium* (L.)]	2839	FDA §172.510	Natural flavoring substances and natural substances used in conjunction with flavors
Pentaacetylglucose	See glucose pentaacetate			
1,2,3,4,6-Pentaacetyl-*α*-*d*-glucose	See glucose pentaacetate			
1,2,3,4,6-Pentaacetyl-*β*-*d*-glucose	See glucose pentaacetate			
α-Pentaacetyl-*dextro*-glucose	See glucose pentaacetate			
β-Pentaacetyl-*dextro*-glucose	See glucose pentaacetate			
ω-Pentadecalactone	Angelicalactone; 15-hydroxypentadecanoic acid, *ω*-lactone; pentadecanolide	2840	FDA §172.515	Synthetic flavor
Pentadecanolide	See *ω*-pentadecalactone			
2,4-Pentadienal		3217		
Pentaerythritol ester of fumaric acid-rosin adduct			FDA §8.390	Diluent in mixtures for coloring egg shells; no penetration into egg
Pentaerythritol ester of partially hydrogenated gum or wood rosin			FDA §121.1059	Plasticizing material in chewing-gum base
Pentaerythritol ester of wood rosin			FDA §121.1059	Plasticizing material in chewing-gum base; product specifications apply
Pentanal	See valeraldehyde			
2,3-Pentanedione	Acetyl propionyl	2841	FDA §172.515	Synthetic flavor
Pentanoic acid	See valeric acid			
1-Pentanol	See amyl alcohol			
2-Pentanol		3316		See *Food Technol.*, May 1972
Pentanolide-1,4	See *γ*-valerolactone			
2-Pentanone	Methyl propyl ketone	2842	FDA §172.515	Synthetic flavor

REGULATORY STATUS OF DIRECT FOOD ADDITIVES (continued)

Material	Alternate Name [Source]	FEMA No.	Regulation	Limitations
2-Pentenal		3218		
4-Pentenoic acid	Allyl acetic acid	2843	FDA §172.515	Synthetic flavor
1-Penten-3-ol	Ethyl vinyl carbinol	3584	FDA §172.515	Synthetic flavor/adjuvant; see *Food Technol.*, February 1978
2-Penten-1-ol		—		
1-Penten-3-one		3382		
3-Penten-2-one		3417		
Pentyl (iso) acetoacetate	See isoamyl acetoacetate			
Pentyl alcohol	See amyl alcohol			
Pentyl (iso) alcohol	See isoamyl alcohol			
Pentyl (iso) amine	See isopentylamine			
Pentyl (iso) benzoate	See isoamyl benzoate			
Pentyl (iso) butanoate	See isoamyl butyrate			
Pentyl butyrate	See amyl butyrate			
Pentyl (iso) butyrate	See isoamyl butyrate			
Pentyl (iso) caproate	See isoamyl hexanoate			
α-Pentylcinnamaldehyde	See α-amylcinnamaldehyde			
α-Pentylcinnamaldehyde, dimethyl acetal	See α-amylcinnamaldehyde dimethyl acetal			
Pentyl (iso) cinnamate	See isoamyl cinnamate			
α-Pentylcinnamyl acetate	See α-amylcinnamyl acetate			
α-Pentylcinnamyl alcohol	See α-amylcinnamyl alcohol			
α-Pentylcinnamyl formate	See α-amylcinnamyl formate			
α-Pentylcinnamyl isovalerate	See α-amylcinnamyl isovalerate			
Pentyl crotonyl acetate	See 1-octen-3-yl acetate			
Pentyl decanoate	See amyl decanoate			
Pentyl (iso) dodecanoate	See isoamyl laurate			
Pentyl formate	See amyl formate			
Pentyl (iso) formate	See isoamyl formate			
2-Pentyl furan		3317		See *Food Technol.*, May 1972
Pentyl (iso) 2-furanbutyrate	See isoamyl 2-furanbutyrate			

Pentyl (iso) 2-furanpropionate	See isoamyl 2-furanpropionate	
Pentyl 2-furoate	See amyl 2-furoate	
Pentyl-2-furyl ketone		3418
Pentyl heptanoate	See amyl heptanoate	
Pentyl hexanoate	See amyl hexanoate	
Pentyl (iso) hexanoate	See isoamyl hexanoate	
Pentyl 2-hydroxybenzoate	See amyl salicylate	
Pentyl (iso) *O*-hydroxybenzoate	See isoamyl salicylate	
Pentyl (iso) isobutyrate	See isoamyl isobutyrate	
Pentyl (iso) isopentanoate	See isoamyl isovalerate	
Pentyl (iso) isovalerate	See isoamyl isovalerate	
Pentyl (iso) β-ketobutyrate	See isoamyl acetoacetate	
Pentyl (iso) laurate	See isoamyl laurate	
n-Pentyl methanoate	See amyl formate	
Pentyl (iso) methanoate	See isoamyl formate	
Pentyl 3-methylbutanoate	See amyl isovalerate	
Pentyl (iso) 3-methylbutanoate	See isoamyl isovalerate	
Pentyl (iso) 2-methylbutyrate	See isoamyl 2-methylbutyrate	
Pentyl (iso) 2-methylpropanoate	See isoamyl isobutyrate	
Pentyl (iso) nonanoate	See isoamyl nonanoate	
Pentyl octanoate	See amyl octanoate	
Pentyl (iso) octanoate	See isoamyl octanoate	
Pentyl octylate	See amyl octanoate	
Pentyl (iso) octylate	See isoamyl octanoate	
Pentyl (iso) oxobutanoate	See isoamyl acetoacetate	
Pentyl 2-phenylacetate	See amyl phenylacetate	
Pentyl (iso) phenylacetate	See isoamyl phenylacetate	
α-Pentyl-β-phenylacrolein	See α-amylcinnamaldehyde	
Pentyl (iso) 3-phenylacrylate	See isoamyl cinnamate	
Pentyl (iso) 3-phenylpropenoate	See isoamyl cinnamate	
Pentyl (iso) propanoate	See isoamyl propionate	

REGULATORY STATUS OF DIRECT FOOD ADDITIVES (continued)

Material	Alternate Name [Source]	FEMA No.	Regulation	Limitations
Pentyl (iso) propionate	See isoamyl propionate			
2-Pentylpyridine		3383		
Pentyl (iso) pyruvate	See isoamyl pyruvate			
Pentyl- (iso) salicylate	See isoamyl salicylate			
Pentyl α-toluate	See amyl phenylacetate			
Pentyl vinyl carbinol	See 1-octen-3-ol			
n-Pentyl vinyl carbinol acetate	See 1-octen-3-yl acetate			
Pepper, black	[*Piper nigrum* (L.)]	2844	FDA §182.1	GRAS, natural flavor and natural flavor extractive
Pepper, black, oil	[*P. nigrum* (L.)]	2845	FDA §182.1	Natural flavor
Pepper, black, oleoresin	[*P. nigrum* (L.)]	2846	FDA §182.1	Natural flavor
Pepper, cayenne	See capsicum [*Capsicum frutescens* (L.) of *C. annuum* (L.)]			
Peppermint camphor	See menthol			
Peppermint, leaves	[*Mentha piperita* (L.)]	2847		
Peppermint, oil	[*M. piperita* (L.)]	2848	FDA §182.1	GRAS, natural flavor and natural flavor extractive
Pepper, red	[*C. frutescens* (L.) or *C. annuum* (L.)]	2849	FDA §182.1	Natural flavor
Pepper tree, oil	See *Schinus molle*, oil [*Schinus molle* (L.)]			
Pepper, white	[*P. nigrum* (L.)]	2850	FDA §182.1	GRAS, natural flavor and natural flavor extractive
Pepper, white, oil	[*P. nigrum* (L.)]	2851	FDA §182.1	Natural flavor
Pepper, white, oleoresin	[*P. nigrum* (L.)]	2852	FDA §182.1	Natural flavor
Perilla alcohol	See *p*-mentha-1,8-dien-7-ol			
Perillaldehyde	4-Isopropyl-1-cyclohexene-1-carboxaldehyde; *p*-mentha-1,8-diene-7-al	—	FDA §172.510	Synthetic flavor
Perillo	[*Couma macrocarpa* Barb. and *C. utilis* (Mart.) Muell. Arg.]		FDA §121.1059	Natural (coagulated or concentrated latex) masticatory substance in chewing-gum base
Perillyl acetate	*p*-Mentha-1,8-dien-7-yl acetate	—	FDA §172.515	Synthetic flavor

Perillyl alcohol	See *p*-mentha-1,8-dien-7-ol			
Persic oil	See apricot kernel [*Prunus armenica* (L.)] and peach kernel [*P. persica* Sieb and Zucc.]			
Peruvian balsam	Balsam of Peru [*Myroxylon pereirae* Klotzsch]		FDA §182.1	GRAS, natural flavor extractive
Peruviol	See nerolidol			
Petitgrain lemon, oil	[*Citrus limonum* (L.) Burm. f.]	2853	FDA §182.1	GRAS, natural flavor extractive
Petitgrain mandarin oil	[*C. reticulate* Blanco var. *mandarin*]	2854	FDA §182.1	Natural flavor
Petitgrain mandarin or tangerine	See mandarin, oil [*C. reticulata* Blanco var. *mandarin*]			
Petitgrain, oil	[*C. aurantium* (L.)]	2855	FDA §182.1	GRAS, natural flavor extractive
Petrohol®	See isopropyl alcohol			
Petrolatum	Paraffin jelly		FDA §121.1166	1. *0.15%* max in bakery products, as release agent and lubricant with white mineral oil 2. 0.02% max in dehydrated fruits and vegetables, as release agent 3. 0.10% in egg-white solids, as release agent 4. On raw fruits and vegetables, as protective coating 5. Defoaming agent in beet sugar and yeast, 150 ppm max in yeast, measured as hydrocarbons 6. *Conforming* with §121.261
			FDA §121.1099	In defoaming agents limited in use to processing beet sugar and yeast, *150* ppm max in yeast, measured as hydrocarbons; product specifications apply
Petrolatum, liquid	See mineral oil			
Petroleum wax			FDA §121.1156	1. Masticatory substance in chewing-gum base 2. Protective coating on cheese, raw fruits, and raw vegetables 3. Defoamer in food, in accordance with §121.1099

REGULATORY STATUS OF DIRECT FOOD ADDITIVES (continued)

Material	Alternate Name [Source]	FEMA No.	Regulation	Limitations
				4. Product specifications apply; may contain any antioxidant permitted in food, in accordance with §409
			FDA §121.1099	In defoaming agents, limited in use to processing beet sugar and yeast; 150 ppm max in yeast, measured as hydrocarbons
α-Phellandrene	*p*-Mentha-1,5-diene	2856	FDA §172.515	Synthetic flavor
1-Phenethoxy-1-propoxyethane	See acetaldehyde phenethyl propyl acetal			
Phenethyl acetate	2-Phenylethyl acetate	2857	FDA §172.515	Synthetic flavor
Phenethyl alcohol	β-Phenylethyl alcohol	2858	FDA §172.515	Synthetic flavor
Phenethylamine		3220		
Phenethyl anthranilate	2-Phenylethyl anthranilate	2859	FDA §172.515	Synthetic flavor
Phenethyl benzoate	2-Phenylethyl benzoate	2860	FDA §172.515	Synthetic flavor
Phenethyl butyrate	2-Phenylethyl butyrate	2861	FDA §172.515	Synthetic flavor
Phenethyl carbinol	See 1-Phenyl-1-propanol			
Phenethyl cinnamate	2-Phenylethyl cinnamate	2863	FDA §172.515	Synthetic flavor
Phenethyl 3,3-dimethylacrylate	See phenethyl senecioate			
Phenethyl formate	2-Phenylethyl formate	2864	FDA §172.515	Synthetic flavor
Phenethyl 2-furoate	2-Phenylethyl 2-furoate	2865		
Phenethyl hexanoate		3221		
Phenethyl isobutyrate	2-Phenylethyl isobutyrate	2862	FDA §172.515	Synthetic flavor
Phenethyl isovalerate	2-Phenylethyl isovalerate	2871	FDA §172.515	Synthetic flavor
Phenethyl isovalerianate	See phenethyl isovalerate			
β-Phenethyl α-methylbutanoate	See phenethyl 2-methylbutyrate			
Phenethyl 2-methylbutyrate		—	FDA §172.515	Synthetic flavor/adjuvant
Phenethyl 3-methylbutyrate	See phenethyl isovalerate			
Phenethyl 3-methylcrotonate	See phenethyl senecioate			

Phenethyl methyl ether	See methyl phenethyl ether			
Phenethyl 2-methylpropanoate	See phenethyl isobutyrate			
Phenethyl octanoate		3222		
Phenethyl phenylacetate	2-Phenylethyl phenylacetate	2866	FDA §172.515	Synthetic flavor
β-Phenethyl β-phenylacrylate	See phenethyl cinnamate			
Phenethyl propionate	2-Phenylethyl propionate	2867	FDA §172.515	Synthetic flavor
Phenethyl salicylate	2-Phenylethyl salicylate	2868	FDA §172.515	Synthetic flavor
Phenethyl senecioate	Phenethyl 3,3-dimethylacrylate	2869	FDA §172.515	Synthetic flavor
Phenethyl tiglate	2-Phenylethyl tiglate	2870	FDA §172.515	Synthetic flavor
Phenol		3223		0.50 ppm max in beverages, frozen desserts, candy, baked goods, gelatins, and puddings
Phenoxyacetic acid	Phenoxyethanoic acid	2872	FDA §172.515	Synthetic flavor
Phenoxyethanoic acid	See phenoxyacetic acid			
2-Phenoxyethyl isobutyrate		2873	FDA §172.515	Synthetic flavor
2-Phenoxyethyl 2-methylpropanoate	See 2-phenoxyethyl isobutyrate			
Phenylacetaldehyde	α-Toluic aldehyde	2874	FDA §172.515	Synthetic flavor
Phenylacetaldehyde 2,3-butylene glycol acetal		2875	FDA §172.510	Synthetic flavor
Phenylacetaldehyde diisobutyl acetal		3384		
Phenylacetaldehyde dimethyl acetal	Viridine	2876	FDA §172.510	Synthetic flavor
Phenylacetaldehyde glyceryl acetal		2877	FDA §172.510	Synthetic flavor
Phenylacetic acid	α-Toluic acid	2878	FDA §172.510	Synthetic flavor
Phenylacetic acid, benzyl ester	See benzyl phenylacetate			
Phenylacetic acid, hexyl ester	See hexyl phenylacetate			
Phenylacetic acid, isopentyl ester	See isoamyl phenylacetate			
Phenylacetic acid, p-methoxybenzyl ester	See anisyl phenylacetate			
Phenylacetic aldehyde	See phenylacetaldehyde			

REGULATORY STATUS OF DIRECT FOOD ADDITIVES (continued)

Material	Alternate Name [Source]	FEMA No.	Regulation	Limitations
Phenylacrolein	See cinnamaldehyde			
β-Phenylacrolein	See cinnamaldehyde			
3-Phenylacrylic acid	See cinnamic acid			
β-Phenylacrylic acid	See cinnamic acid			
Phenylalanine (L-and DL-forms)			FDA §182.1	GRAS, nutrient and/or dietary supplement, as the free acid, hydrochloride salt, hydrate, or anhydrous form
3-Phenylallyl acetate	See cinnamyl acetate			
γ-Phenylallyl alcohol	See cinnamyl alcohol			
3-Phenylallyl butyrate	See cinnamyl butyrate			
Phenylallyl cinnamate	See cinnamyl cinnamate			
3-Phenylallyl cinnamate	See cinnamyl cinnamate			
3-Phenylallyl formate	See cinnamyl formate			
3-Phenylallyl isovalerate	See cinnamyl isovalerate			
3-Phenylallyl 3-methylbutanoate	See cinnamyl isovalerate			
3-Phenylallyl phenylacetate	See cinnamyl phenylacetate			
3-Phenylallyl propionate	See cinnamyl propionate			
γ-Phenylallyl propionate	See cinnamyl propionate			
Phenylamyl alcohol	See 5-phenylpentanol			
Phenylbenzene	See biphenyl			
4-Phenyl-2-butanol	Phenylethyl methyl carbinol	2879	FDA §172.515	Synthetic flavor
2-Phenyl-2-butenal		3224		
4-Phenyl-3-buten-2-ol	Methyl styryl carbinol	2880	FDA §172.515	Synthetic flavor
4-Phenyl-3-buten-2-one	Benzilidene acetone	2881	FDA §172.515	Synthetic flavor
4-Phenyl-2-butyl acetate	Phenylethyl methyl carbinyl acetate	2882	FDA §172.515	Synthetic flavor
2-Phenyl-3-carbethoxy furan		3468		
Phenyl carbinol	See benzyl alcohol			
Phenyl carboxylic acid	See benzoic acid			
Phenylcellosolve isobutyrate	See 2-phenoxyethyl isobutyrate			

2-Phenylcrotonaldehyde	See 2-phenyl-2-butenal	
Phenyl dimethyl carbinyl isobutyrate	See α,α-dimethylbenzyl isobutyrate	
2-Phenyl-*m*-dioxan-5-ol	See benzaldehyde glyceryl acetal	
2-Phenyl-1,3-dioxan-5-ol	See benzaldehyde glyceryl acetal	
Phenyl disulfide	Diphenyl disulfide	3225
Phenyldithiobenzene	See phenyl disulfide	
1-Phenylethanol	See α-methylbenzyl alcohol	
1-Phenylethan-1-ol	See α-methylbenzyl alcohol	
1-Phenyl-2-ethanol	See phenethyl alcohol	
Phenylethene	See styrene	
2-Phenylethyl acetate	See phenethyl acetate	
α-Phenylethyl acetate	See α-methylbenzyl acetate	
sec-Phenylethyl acetate	See α-methylbenzyl acetate	
2-Phenylethyl alcohol	See phenethyl alcohol	
α-Phenylethyl alcohol	See α-methylbenzyl alcohol	
β-Phenylethyl alcohol	See phenethyl alcohol	
2-Phenylethylamine	See phenethylamine	
2-Phenylethyl anthranilate	See phenethyl anthranilate	
2-Phenylethyl benzoate	See phenethyl benzoate	
2-Phenylethyl butanoate	See phenethyl butyrate	
1-Phenyl-1-ethyl butanoate	See α-methylbenzyl butyrate	
2-Phenylethyl butyrate	See phenethyl butyrate	
2-Phenylethyl caproate	See phenethyl hexanoate	
2-Phenylethyl caprylate	See phenethyl octanoate	
Phenylethyl carbinol	See 1-phenyl-1-propanol	
2-Phenylethyl cinnamate	See phenethyl cinnamate	
2-Phenylethyl *trans*-2,3-dimethylacrylate	See phenethyl tiglate	
Phenylethyl dimethyl carbinyl isobutyrate	See 2-methyl-4-phenyl-2-butyl isobutyrate	
Phenylethylene	See styrene	
2-Phenylethyl formate	See phenethyl formate	
1-Phenyl-1-ethyl formate	See α-methylbenzyl formate	
2-Phenylethyl 2-furoate	See phenethyl 2-furoate	
2-Phenylethyl hexanoate	See phenethyl hexanoate	
2-Phenylethyl 2-hydroxybenzoate	See phenethyl salicylate	

REGULATORY STATUS OF DIRECT FOOD ADDITIVES (continued)

Material	Alternate Name [Source]	FEMA No.	Regulation	Limitations
2-Phenylethyl isobutyrate	See phenethyl isobutyrate			
1-Phenyl-1-ethyl isobutyrate	See α-methylbenzyl isobutyrate			
2-Phenylethyl isovalerate	See phenethyl isovalerate			
Phenyl ethyl ketone	See propiophenone			
2-Phenylethyl methanoate	See phenethyl formate			
1-Phenyl-1-ethyl methanoate	See α-methylbenzyl formate			
2-Phenylethyl 2-methylbutanoate	See phenethyl 2-methylbutyrate			
2-Phenylethyl 3-methylbutanoate	See phenethyl isovalerate			
2-Phenylethyl *trans*-2-methylbutanoate	See phenethyl tiglate			
2-Phenylethyl 3-methyl-2-butanoate	See phenethyl senecioate			
Phenylethyl methyl carbinol	See 4-phenyl-2-butanol			
Phenylethyl methyl carbinyl acetate	See 4-phenyl-2-butyl acetate			
Phenylethyl methyl ethyl carbinol	See 1-phenyl-3-methyl-3-pentanol			
1-Phenyl-1-ethyl 2-methylpropanoate	See α-methylbenzyl isobutyrate			
2-Phenylethyl pentanoate	See 2-phenylethyl valerate			
2-Phenylethyl phenylacetate	See phenethyl phenylacetate			
2-Phenylethyl 3-phenylpropanoate	See phenethyl cinnamate			
2-Phenylethyl propanoate	See phenethyl propionate			
2-Phenylethyl propionate	See phenethyl propionate			
1-Phenyl-1-ethyl propionate	See α-methylbenzyl propionate			

2-Phenylethyl salicylate	See phenethyl salicylate			
2-Phenylethyl senecioate	See phenethyl senecioate			
2-Phenylethyl tiglate	See phenethyl tiglate			
2-Phenylethyl α-toluate	See phenethyl phenylacetate			
2-Phenylethyl valerate		—		
Phenylformic acid	See benzoic acid			
2-Phenyl-3-(2-furyl)-prop-2-enal	2-Furfurylidenephenyl-acetaldehyde	3586		See *Food Technol.*, February 1978
o-Phenylglycolic acid	See phenoxyacetic acid			
1-Phenyl-1-hydroxyethane	See α-methylbenzyl alcohol			
α-Phenylisopentanal	See 3-methyl-2-phenylbutyraldehyde			
α-Phenylisovaleraldehyde	See 3-methyl-2-phenylbutyraldehyde			
Phenyl ketone	See benzophenone			
Phenylmethanethiol	See benzyl mercaptan			
Phenylmethanol	See benzyl alcohol			
3-Phenyl-2-methylacrolein	See α-methylcinnamaldehyde			
Phenylmethyl alcohol	See benzyl alcohol			
Phenyl methyl diketone	See 1-phenyl-1,2-propanedione			
Phenyl methyl ether	See anisole			
Phenyl methyl ketone	See acetophenone			
1-Phenyl-3-methyl-3-pentanol	Phenylethyl methyl ethyl carbinol	2883	FDA §172.515	Synthetic flavor
5-Phenylpentanol		—		
1-Phenyl-2-pentanol	See α-propylphenethyl alcohol			
2-Phenyl-4-pentenal		3519		
3-Phenyl-4-pentenal		3318		See *Food Technol.*, May 1972
2-Phenylpropanal	See 2-phenylpropionaldehyde			
3-Phenylpropanal	See 3-phenylpropionaldehyde			
1-Phenyl-1,2-propanedione		3226		
1-Phenyl-1-propanol	Phenethyl carbinol	2884	FDA §172.515	Synthetic flavor
2-Phenyl-1-propanol	See β-methylphenethyl alcohol			
3-Phenyl-1-propanol	Hydrocinnamyl alcohol	2885	FDA §172.515	Synthetic flavor
1-Phenyl-1-propanone	See propiophenone			
2-Phenylpropan-2-yl isobutyrate	See α,α-dimethylbenzyl isobutyrate			
2-Phenylpropan-2-yl 2-methylpropanoate	See α,α-dimethylbenzyl isobutyrate			
3-Phenylpropenal	See cinnamaldehyde			

REGULATORY STATUS OF DIRECT FOOD ADDITIVES (continued)

Material	Alternate Name [Source]	FEMA No.	Regulation	Limitations
3-Phenyl-2-propen-1-al	See cinnamaldehyde			
3-Phenylpropenoic acid	See cinnamic acid			
3-Phenyl-2-propen-1-ol	See cinnamyl alcohol			
3-Phenyl-2-propen-1-yl acetate	See cinnamyl acetate			
3-Phenyl-2-propen-1-yl benzoate	See cinnamyl benzoate			
3-Phenyl-2-propen-1-yl butanoate	See cinnamyl butyrate			
3-Phenyl-2-propen-1-yl formate	See cinnamyl formate			
3-Phenyl-2-propen-1-yl isobutyrate	See cinnamyl isobutyrate			
3-Phenyl-2-propen-1-yl 3-methylbutanoate	See cinnamyl isovalerate			
3-Phenyl-2-propen-1-yl 2-methyl propanoate	See cinnamyl isobutyrate			
3-Phenyl-2-propen-1-yl phenylacetate	See cinnamyl phenylacetate			
3-Phenyl-2-propen-1-yl 3-phenylpropenoate	See cinnamyl cinnamate			
3-Phenyl-2-propenyl propanoate	See cinnamyl propionate			
3-Phenyl-2-propen-1-yl propionate	See cinnamyl propionate			
2-Phenylpropionaldehyde	Hydratropaldehyde	2886	FDA §172.510	Synthetic flavor
3-Phenylpropionaldehyde	Hydrocinnamaldehyde	2887	FDA §172.510	Synthetic flavor
α-Phenylpropionaldehyde	See 2-phenylpropionaldehyde			
2-Phenylpropionaldehyde dimethyl acetal	Hydrotropic aldehyde dimethyl acetal	2888	FDA §172.515	Synthetic flavor
3-Phenylpropionic acid	Hydrocinnamic acid	2889	FDA §172.515	Synthetic flavor
β-Phenylpropionic acid	See 3-phenylpropionic acid			

3-Phenylpropyl acetate	Hydrocinnamyl acetate	2890	FDA §172.515	Synthetic flavor
β-Phenylpropyl acetate	See 3-phenylpropyl acetate			
Phenylpropyl alcohol	See 3-phenyl-1-propanol			
α-Phenylpropyl alcohol, butyric ester	See 2-phenylpropyl butyrate			
α-Phenylpropyl alcohol, isobutyric ester	See 2-phenylpropyl isobutyrate			
Phenylpropyl aldehyde	See 3-phenylpropionaldehyde			
2-Phenylpropyl butyrate	β-Methylphenethyl butyrate	2891	FDA §172.515	Synthetic flavor
α-Phenylpropyl butyrate	See α-ethylbenzyl butyrate			
1-Phenyl-1-propyl butyrate	See α-ethylbenzyl butyrate			
1-Phenyl-2-propyl butyrate	See α-methylphenethyl butyrate			
3-Phenylpropyl caproate	See 3-phenylpropyl hexanoate			
3-Phenylpropyl cinnamate	Hydrocinnamyl cinnamate	2894	FDA §172.515	Synthetic flavor
β-Phenylpropyl cinnamate	See 3-phenylpropyl cinnamate			
3-Phenylpropyl formate	Hydrocinnamyl formate	2895	FDA §172.515	Synthetic flavor
β-Phenylpropyl formate	See 3-phenylpropyl formate			
3-Phenylpropyl hexanoate	Hydrocinnamyl hexanoate	2896	FDA §172.515	Synthetic flavor
2-Phenylpropyl isobutyrate	Hydratropyl isobutyrate	2892	FDA §172.515	Synthetic flavor
3-Phenylpropyl isobutyrate	Hydrocinnamyl isobutyrate	2893	FDA §172.515	Synthetic flavor
3-Phenylpropyl isopentanoate	See 3-phenylpropyl isovalerate		—	
3-Phenylpropyl isovalerate	Hydrocinnamyl isovalerate	2899	FDA §172.515	Synthetic flavor
3-Phenylpropyl isovalerianate	See 3-phenylpropyl isovalerate			
3-Phenylpropyl methanoate	See 3-phenylpropyl formate			
3-Phenylpropyl 3-methylbutanoate	See 3-phenylpropyl isovalerate			
β-Phenylpropyl 3-methylbutanoate	See 3-phenylpropyl isovalerate			
2-Phenylpropyl 2-methylpropanoate	See 2-phenylpropyl isobutyrate			
3-Phenylpropyl 2-methylpropanoate	See 3-phenylpropyl isobutyrate			
β-Phenylpropyl 2-methylpropanoate	See 3-phenylpropyl isobutyrate			
3-Phenylpropyl β-phenylacrylate	See 3-phenylpropyl cinnamate			

REGULATORY STATUS OF DIRECT FOOD ADDITIVES (continued)

Material	Alternate Name [Source]	FEMA No.	Regulation	Limitations
3-Phenylpropyl 3-phenyl-propenoate	See 3-phenylpropyl cinnamate			
3-Phenylpropyl propanoate	See 3-phenylpropyl propionate			
3-Phenylpropyl propionate	Hydrocinnamyl propionate	2897	FDA §172.515	Synthetic flavor
β-Phenylpropyl propionate	See 3-phenylpropyl propionate			
1-Phenyl-3-or 5-propylpyr-azol		—		
2-(3-Phenylpropyl)-tetrahy-drofuran	2-Hydrocinnamyl tetrahydrofuran	2898	FDA §172.515	Synthetic flavor
3-Phenyl-3-vinylpropional-dehyde	See 3-phenyl-4-pentenal			
Phosphoric acid	Orthophosphoric acid	2900	FDA §182.1	GRAS, miscellaneous and/or general-purpose food additive
			FDA §8.303	In production of caramel
			MID	0.01% to increase effectiveness of an-tioxidants in lard and shortening
Phosphorus oxychloride	Phosphoryl chloride		FDA §121.1031	0.1% in starch-modified food, alone as esterifier or in combination with up to 10% propylene oxide for esterification and etherification
Pimenta leaf, oil (see also allspice)	[*Pimenta officinalis* Lindl.]	2901	FDA §182.1	
Pinanethiol	See 2-, 3-, or 10-mercaptopinane			
Pinanyl mercaptan	See 2-, 3-, or 10-mercaptopinane			
Pine, dwarf, needles and twigs, oil	[*Pinus mugo* Turra var. *pumilio* (Haenke) Zenari]	2904	FDA §172.510	Natural flavor
"Pine" (fir), needles and twigs	See fir ("Pine"), needles and twigs [*Abies* spp.]		FDA §172.515	Synthetic flavor/adjuvant
Pine, mountain, oil	See pine, dwarf, needles and twigs, oil [*Pinus mugo* Turra var. *pumilio* (Haenke) Zenari]			
2(10)-Pinen-3-ol	6,6-Dimethyl-3-hydroxy-2-methylene-bicyclo [3.1.1] heptane pinocarveol	3587		See *Food Technol.*, February 1978

		3587		See *Food Technol.*, February 1978
2-Pinen-4-ol	See verbenol			
Pinene	See α-pinene			
2-Pinene	See α-pinene			
2(10)-Pinene	See β-pinene			
α-Pinene	2-Pinene	2902	FDA §121.1059	In synthetic terpene resins for chewing-gum base; product specifications apply
			FDA §172.515	Synthetic flavor/adjuvant
β-Pinene	2(10)-Pinene	2903	FDA §121.1059	In synthetic terpene resins for chewing-gum base; product specifications apply
Pine, needle, oil	[*Abies* spp.]	2905	FDA§172.510	
2-Pinen-4-ol	See verbenol			
2(10)-Pinen-3-ol	See pinocarveol			
Pine, Scotch, needles and twigs, oil	[*Pinus sylvestris* (L.)]	2906	FDA §172.510	Natural flavor
Pine tar, oil	[*P. palustris* and other *Pinus* spp.]	2907	FDA §172.515	Synthetic flavor
Pine, white, bark	[*P. strobus* (L.)]	—	FDA §172.510	
Pine, white, bark, extract solid	[*P. strobus* (L.)]	—	FDA §172.510	Natural flavor in alcoholic beverages only
Pine, white, bark, oil	[*P. strobus* (L.)]	—	FDA §172.510	
Pine, white, oil	[*P. palustris* Mill. and other *Pinus* spp.]	—	FDA §172.510	Natural flavor
Pinocarveol	See 2(10)-pinen-3-ol	—	FDA §172.515	Synthetic flavor
Pinus pumilio, oil	See pine, dwarf, needles and twigs, oil [*P. mugo* Turra var. *pumilio* (Haenke) Zenari]			
Piperidene	Hexahydropyridine	2908	FDA §172.515	Synthetic flavor
Piperine	Piperolypiperidine	2909	FDA §172.515	Synthetic flavor
Piperitenone	See p-mentha-1,4(8)-dien-3-one	—	FDA §172.515	Synthetic flavor
Piperitenone oxide		—	FDA §172.515	Synthetic flavor
Piperitol	See *p*-menth-1-en-3-ol			
d-Piperitone	*p*-menth-1-en-3one	2910	FDA §172.515	Synthetic flavor
α-Piperitone	See *d*-piperitone			
Piperonal	Heliotropine; 3,4-methyl-enedioxy-benzaldehyde	2911	FDA §182.1	GRAS, synthetic flavor and adjuvant
Piperonyl acetate	Heliotropyl acetate	2912	FDA §172.515	Synthetic flavor
Piperonyl acetone	See 4-(3,4-methylenedioxyphenyl)-2-butanone			

REGULATORY STATUS OF DIRECT FOOD ADDITIVES (continued)

Material	Alternate Name [Source]	FEMA No.	Regulation	Limitations
Piperonyl aldehyde	See piperonal			
Piperonyl isobutyrate		2913	FDA §172.515	Synthetic flavor
Piperonyl 2-methylpropanoate	See piperonyl isobutyrate			
Piperolypiperidine	See piperine			
Pipsissewa leaves, extract	[*Chimaphila umbellata* (L.) Nutt.]	2914	FDA §182.1	GRAS, natural flavor extractive
Plantarome	See yucca, Mohave, extract [*Yucca schidigera* Roezl ex Ortiges (*Y. mohavensis* Sarg.)]			
P.O.A.	See phenoxyacetic acid			
Polyethylene			FDA §121.1059	Synthetic masticatory substance in chewing-gum base; mol wt, 2000-21,000
Polyethylene glycol	Carbowax®; PEG		FDA §121.1179	Coating on fresh citrus fruit; dispersing and defoaming adjuvant, in compliance with FDA §121.1185; averate mol wt, 200-9500
			FDA §121.1185	1. Coating, binder, plasticizing agent, and/or lubricant in tablets used for food 2. Adjuvant and bodying agent in non-nutritive sweeteners listed as GRAS 3. Adjuvant to disperse vitamins and/or mineral preparations 4. Coating on sodium nitrite to inhibit hygroscopicity; product specifications apply
Polyethylene glycol 6000			FDA §121.1057	1. Binder and plasticizing agent in tablets used for food 2. Adjuvant in tablet coatings to improve resistance to moisture and oxidation

				3. Bodying agent with nonnutritive sweeteners (cyclamate and saccharin salts) listed by FDA as GRAS 4. Product specifications apply
Polyglycerol esters of fatty acids			FDA §12.1120	Up to and including the decaglycerol esters of specified fatty acids: 1. Emulsifiers; cloud inhibitors in hibitors in vegetable and salad oils 2. Esters of butter oil fatty acids as emulsifiers in dry, whipped-topping base. 3. Product specifications apply
			FDA §8.303	Antifoaming agents in caramel; conforming with §121.1120
			MID	In rendered animal fat or a combination of animal fat with vegetable fat; product specifications apply
Polylimonene		—	FDA §172.515	Synthetic flavor
Polyoxyethylene (20) sorbitan monooleate	See polysorbate 20			
Polyoxethylene (20) sorbitan 80	See polysorbate 80			
Polyoxyethylene (20) sorbitan monostearate	See Polysorbate 60			
Polyoxyethylene (20) sorbitan tristearate	See Polysorbate 65			
Polysorbate 20	Polyoxyethylene (20) sorbitan monolaurate	2915	FDA §172.515	Synthetic flavor
			FDA §121.2541	Emulsifier and/or surface-active agent; product specifications apply
Polysorbate 60	Polyoxyethylene (20) sorbitan monostearate	2916	FDA §121.1030	1. Emulsifier in whipped vegetable oil topping, alone or with sorbitan monostearate, polyoxyethylene (20) sorbitan tristearate, and/or polysorbate 80; additive(s), 0.40% max 2. Emulsifier in cake and cake mixes, 0.46% max alone; with polyoxyethylene (20) sorbitan tristearate and/or

REGULATORY STATUS OF DIRECT FOOD ADDITIVES (continued)

Material	Alternate Name [Source]	FEMA No.	Regulation	Limitations
				sorbitan monostearate, 0.46% max; polyoxyethylene (20) sorbitan tristearate, 0.32% max; sorbitan monostearate, 0.61% max; total additives, 0.66% max
				3. Emulsifier in shortenings and edible oils for use in nonstandardized baked goods, baking mixes, icings, fillings, and toppings, and in frying of foods; 1.0% alone; with polysorbate 80, total additives, 1.0% max by weight of finished shortening or oil
				4. Emulsifier in cake icings and cake fillings; 0.46% max alone; with polyoxyethylene (20) sorbitan tristearate and/or sorbitan monostearate, 0.46% max; polyoxyethylene (20) sorbitan tristearate, 0.32% max; sorbitan monostearate, 0.70% max; total additives, 1.0% max
				5. To impart greater opacity to sugar-type confection coatings, 0.2% max
				6. Emulsifier in nonstandardized dressings made without egg yolk, 0.3% max
				7. Emulsifier in solid-state, edible vegetable-fat-water emulsions for use as substitutes for milk or cream in beverage coffee; 0.4% max, alone or in combination with polyoxyethylene (20) sorbitan tristearate and/or sorbitan monostearate
				8. Foaming agent in nonalcoholic mixes, to be added to alcoholic bever-

			ages in preparation of mixed drinks; 4.5% max by weight of the nonalcoholic mix Product specifications apply
		FDA §172.515	Synthetic flavor
		MID	To emulsify shortenings sold in units not exceeding 6 lb or 1 gal fluid content; 1% max alone; with polysorbate 80 or sorbitan monostearate, total additives, 1% max
Polysorbate 65	Polyoxyethylene (20) sorbitan tristearate	FDA §121.1008	1. Emulsifier in ice cream, frozen custard, ice milk, fruit sherbet, and nonstandardized frozen desserts, alone or with polysorbate 80; total additive(s), 0.10% max 2. Emulsifier in cakes and cake mixes, 0.32% max alone; with sorbitan monostearate and/or polysorbate 60, 0.32% max; sorbitan monostearate, 0.61% max; polysorbate 60, 0.46% max; total additives; 0.66% max 3. Emulsifier in whipped vegetable oil topping, alone or with sorbitan monostearate, polysorbate 60, and/or polysorbate 80; total additives, 0.40% max 4. Emulsifier in cake icings and cake fillings, 0.32% max alone; with sorbitan monostearate and/or polysorbate 60, 0.32% max; sorbitan monostearate, 0.70% max; polysorbate 60, 0.46% max; total additives, 1.0% max 5. Emulsifier in solid-state, edible vegetable-fat-water emulsions for use as substitutes for milk or cream in beverage coffee — 0.4% max, alone or with polysorbate 60 and/or sorbitan monostearate; product specifications apply

REGULATORY STATUS OF DIRECT FOOD ADDITIVES (continued)

Material	Alternate Name [Source]	FEMA No.	Regulation	Limitations
			FDA §121.1099	Defoaming agent
			FDA §121.2541	Emulsifier and/or surface-active agent
Polysorbate 80	Polyoxyethylene (20) sorbitan monooleate	2917	FDA §121.1009	1. Emulsifier in ice cream, frozen custard, ice milk, fruit sherbet, and nonstandardized frozen desserts, alone or with polyoxyethylene (20) sorbitan tristearate; total additive(s), 0.10% max 2. Emulsifier in shortenings and edible oils for use in nonstandardized baked goods, baking mixes, icings, fillings, and toppings, and in the frying of foods, alone, 1.0% max; with polysorbate 60, total additives, 1.0% max 3. In yeast-defoamer formulations, 4% max of the finished yeast defoamer or of the yeast from such use 4. Solubilizing and dispersing agent in pickles and pickle products, up to 0.05% 5. Solubilizing and dispersing agent for fat-soluble vitamins in vitamin and vitamin-mineral preparations; intake from recommended daily dose, 300 mg max; in vitamin A preparations containing more than 30,000 units per dose, daily intake, 500 mg max 6. Surfactant in production of coarse crystal sodium chloride, up to 10 ppm in finished salt 7. Emulsifier for edible fats and oils in special dietary foods, daily ingestion, not to exceed 360 mg

			8. Emulsifier in whipped, vegetable oil topping, alone or with sorbitan monostearate, polysorbate 60, and/or polyoxyethylene (20) sorbitan tristearate; total additive(s), 0.4%
			9. Wetting agent in scald water for poultry defeathering, following by potable water rinse; 0.0175% max in scald water.
			10. Solubilizing and dispersing agent for dill oil in canned, spiced green beans, up to 0.003%
			11. Dispersing agent in gelatin desserts and in gelatin dessert mixes, up to 0.082%
			Product specifications apply
		MID	To emulsify shortenings sold in units not exceeding 6 lb or 1 gal fluid content, 1% alone; with polysorbate 60 or sorbitan monostearate, total additives, 1% max
Polyvinyl acetate		FDA §8.390	Diluent in inks for marking gum, confectionery, and food supplements in tablet form; min mol wt, 2000
Polyvinyl alcohol	Polyviol®	FDA §8.390	Diluent in mixtures for coloring egg shells; no penetration into egg
Polyvinylpyrrolidone	Kollidon®	FDA §121.1110	Clarifying agent in beverages and vinegar, followed by removal with filtration; product specifications apply
		FDA §121.1139	1. 10 ppm max residue in beer; clarifying agent
			2. 40 ppm max residue in vinegar; clarifying agent
			3. 60 ppm max residue in wine; clarifying agent
			4. Tableting adjuvant in flavor concentrates, nonnutritive sweeteners, and vitamin and mineral concentrates

REGULATORY STATUS OF DIRECT FOOD ADDITIVES (continued)

Material	Alternate Name [Source]	FEMA No.	Regulation	Limitations
				5. Stabilizer, bodying agent, and dispersant in nonnutritive sweeteners in concentrated liquid form and in vitamin and mineral concentrates in liquid form
				Product specifications apply
			FDA §8.300 and §8.6200	Diluent in color additive mixtures in or on food-tablet coatings at 0.1% of finished food; inks for marking gum, confectionery, and food supplements in tablet form; inks for marking fruits and vegetables; inks for branding pharmaceutical forms; and ingested drugs; specifications in §121.1139 appy
Pomegranate, bark, extract	[*Punica granatum* (L.)	2918	FDA §182.1	GRAS, natural flavor extractive
Poplar buds	Aspen [*Populus balsamifera* (L.) (*P. tacamahacca* Mill.), *P. candicans* Alit., or *P. nigra* (L.)]	—	FDA §172.510	Natural flavor in alcoholic beverages only
Poppy seed	[*Papaver somniferum* (L.); *P. sativum* Fuchs]	2919	FDA §182.1	GRAS, natural flavor
Potassium acetate		2920	FDA §172.515	Synthetic flavor
Potassium acid tartrate	See cream of tartar			
Potassium alginate			FDA §182.1	GRAS, stabilizer
Potassium alum			FDA §19	In specified cheeses, up to 6 times the weight of benzoyl peroxide used
			FDA §15	Bleaching agent in flour, not more than 6 parts by weight, alone or with calcium sulfate, magnesium carbonate, sodium aluminum sulfate, dicalcium phosphate, tricalcium phosphate, starch, and/or calcium carbonate and with 1 part by weight of benzoyl peroxide

Potassium bicarbonate	Potassium acid carbonate	FDA §182.1	GRAS, miscellaneous and/or general-purpose food additive
Potassium bisulfite	Potassium hydrosulfide	FDA §182.1	GRAS, chemical preservative; not for use in meats or in foods recognized as source of vitamin B_1
Potassium bromate		FDA §15.20 and §15.30	50 ppm max in unfinished, bromated flour and enriched, bromated flour
		FDA §15.90	75 ppm max in whole-wheat flour
		FDA §17	0.0075% max in bakery products, alone or with potassium iodate, calcium iodate, and/or calcium peroxide
		FDA §121.1020	In fermented malt beverages and distilled spirits made from potassium bromate-treated malt
		FDA §121.1194	In malting of barley, 75 ppm max in malt (calculated as Br); treated malt used only in production of fermented malt beverages or distilled spirits; total bromide residue in fermented malt beverages, 25 ppm max (calculated as Br); product specifications apply
Potassium carbonate	Salt of tartar	FDA §182.1	GRAS, miscellaneous and/or general-purpose food additive
		FDA §8.303	In production of caramel; includes bicarbonate
Potassium caseinate		FDA §20	In frozen desserts
Potassium chloride		FDA §29.4 and §29.5	In artificially sweetened fruit jellies and preserves
		FDA §182.1	GRAS, nutrient and/or dietary supplement
Potassium citrate		FDA §182.1	GRAS, sequestrant
		FDA §182.1	GRAS, miscellaneous and/or general-purpose food additive
Potassium gibberellate		FDA §121.1010	In malting of barley; total with gibberellic acid, 2 ppm max (expressed as gibberellic acid); treated malt to be used only in production of fermented malt beverages or distilled spirits; finished

REGULATORY STATUS OF DIRECT FOOD ADDITIVES (continued)

Material	Alternate Name [Source]	FEMA No.	Regulation	Limitations
				distilled spirits contain none and finished malt beverage contains 5 ppm max of gibberellic acid; product specifications apply
Potassium glycerophosphate			FDA §182.1	GRAS, nutrient and/or dietary supplement
Potassium hydroxide	Potassium hydrate		FDA §182.1	GRAS, miscellaneous and/or general-purpose food additive
			FDA §8.303	In production of caramel
Potassium iodate			FDA §17	In bakery products, 0.0075% max by weight of flour, alone or with potassium bromate, calcium iodate, and/or calcium peroxide
Potassium iodide			FDA §182.1	GRAS, nutrient and/or dietary supplement in table salt as source of dietary iodine, 0.01% max
			FDA §121.1073	Nutritional supplement in mineral and vitamin-mineral preparations; max daily intake of iodine, 0.15 mg.; product specifications apply
Potassium metabisulfite	Potassium pyrosulfite		FDA §182.1	GRAS, chemical preservative; not for use in meats or in foods recognized as source of vitamin B_1
Potassium nitrate			FDA §121.1132	Curing agent in processing of cod roe, 200 ppm max
			MID	Source of nitrite in cured products; 7 lb/100 gal pickle; 3½ oz/100 lb meat (dry cure); 2¾ oz/100 lb chopped meat
Potassium nitrite			MID	To fix color in cured products; 2 lb/100 gal pickle at 10% pump level; 1 oz/100 lb meat (dry cure); ¼ oz/100 lb chopped meat and/or meat byproduct; 200 ppm max total nitrite in finished

				product; product specifications apply
Potassium phosphate			FDA §8.303	In production of caramel (including di- and monobasic phosphate)
Potassium salts of fatty acids			FDA §121.1071	Binder, emulsifier, and anticaking agent in food, alone or with other salts of fatty acids; fatty acids must comply with §121.1070
Potassium sorbate	Sorbic acid potassium salt	2921	FDA §182.1	GRAS, chemical preservative
Potassium sulfate	Sal polychrestum		FDA §182.1	GRAS, miscellaneous and/or general-purpose food additive
			FDA §8.303	In production of caramel
Potassium sulfite			FDA §8.303	In production of caramel
Pot marigold	See marigold, pot			
Pot marjoram	See marjoram, pot			
Prickly ash bark	[*Xanthoxylum americanum* Mill.; *X. clavaherculis* (L.)]		FDA §182.1	GRAS, natural flavor extractive
Proline (L- and DL-forms)			FDA §182.1	GRAS, nutrient and/or dietary supplement, as free acid, hydrochloride salt, hydrate, or anhydrous form
L-Proline		3319		See *Food Technol.*, May 1972
Propanal	See propionaldehyde			
Propane	Dimethylmethane		FDA §182.1	GRAS, miscellaneous and/or general-purpose food additive
			FDA §121.1065	Propellant and aerating agent for foamed or sprayed food products with octafluorocyclobutane, except standardized foods that do not provide for such use
			FDA §121.1181	Propellant and aerating agent for foamed or sprayed food products with chloropentafluoroethane
1,2-Propanediol	See propylene glycol			
1,2-Propanediol dibenzoate	See propylene glycol dibenzoate			
1,2-Propanedithiol		3520		
1,3-Propanedithiol	1,3-Dimercaptopropane	3588		See *Food Technol.*, February 1978
Propanethiol		3521	FDA §172.515	
Propanethiol	See propyl mercaptan			
1-Propanethiol	See propanethiol			

REGULATORY STATUS OF DIRECT FOOD ADDITIVES (continued)

Material	Alternate Name [Source]	FEMA No.	Regulation	Limitations
1,2,3-Propanetriol	See glycerol			
1,2,3-Propanetriol tribenzoate	See glyceryl tribenzoate			
Propanoic acid	See propionic acid			
1-Propanol	See propyl alcohol			
2-Propanol	See isopropyl alcohol			
Propanol (iso)	See isopropyl alcohol			
2-Propanone	See acetone			
Propene, 1-(*p*-methoxyphenyl)	See anethole			
1-Propene, 1-(4-methoxyphenyl)	See anethole			
2-Propene-1-thiol	See allyl mercaptan			
1,2,3-Propenetricarboxylic acid	See aconitic acid			
1-Propene-1,2,3-tricarboxylic acid	See aconitic acid			
2-Propenoic acid, 3-phenyl, phenylmethyl ester	See benzyl cinnamate			
p-Propenyl anisole	Anethole		FDA §182.1	GRAS, synthetic flavor and adjuvant
p-Propenyl (iso) benzene	See *p*,α-dimethylstyrene			
4-Propenyl-1-(benzyloxy)-2-methoxybenzene	See isoeugenyl benzyl ether			
2-Propen-1-yl butanoate	See allyl butyrate			
2-Propenyl *trans*-2-butenoate	See allyl crotonate			
Propenyl cinnamate	See allyl cinnamate			
2-Propen-1-yl cyclohexaneacetate	See allyl cyclohexaneacetate			
2-Propen-1-yl cyclohexanebutyrate	See allyl cyclohexanebutyrate			
2-Propen-1-yl cyclohexanecaproate	See allyl cyclohexanehexanoate			

2-Propen-1-yl cyclohexane-pentanoate	See allyl cyclohexanevalerate			
2-Propen-1-yl cyclohexane-propionate	See allyl cyclohexanepropionate			
2-Propen-1-yl cyclohexane-valerate	See allyl cyclohexanevalerate			
1-Propenyl methyl disulfide	See methyl 1-propenyl disulfide			
2-Propenyl disulfide	See allyl disulfide			
3-Propenyl-6-ethoxyphenol	See propenyl guaethol			
2-Propenyl 2-ethylbutan-oate	See allyl 2-ethylbutyrate			
2-Propenyl 2-ethylbutyrate	See allyl 2-ethylbutyrate			
Propenyl guaethol	6-Ethoxy-*m*-anol	2922	FDA §172.515	Synthetic flavor
4-Propenylguaiacol	See iso-eugenol			
4-Propenylguaiacyl pheny-lacetate	See isoeugenyl phenylacetate			
2-Propenyl 2,4-hexadi-enoate	See allyl sorbate			
2-Propenyl hexanoate	See allyl hexanoate			
2-Propenyl isopentanoate	See allyl isovalerate			
2-Propenyl isothiocyanate	See allyl isothiocyanate			
2-Propenyl isovalerate	See allyl isovalerate			
4-(1-Propen-1-yl)-2-meth-oxyphenyl formate	See isoeugenyl formate			
4-(2-Propen-1-yl)-2-meth-oxyphenyl formate	See eugenyl formate			
2-Propenyl 3-methylbutan-oate	See allyl isovalerate			
2-Propenyl nonanoate	See allyl nonanoate			
2-Propenyl octanoate	See allyl octanoate			
2-Propenyl octylate	See allyl octanoate			
2-Propenyl pelargonate	See allyl nonanoate			
2-Propenyl phenoxyacetate	See allyl phenoxyacetate			
2-Propenyl phenylacetate	See allyl phenylacetate			
p-Propenylphenyl methyl ether	See anethole			
2-Propenyl polysulfides	See diallyl polysulfides			

REGULATORY STATUS OF DIRECT FOOD ADDITIVES (continued)

Material	Alternate Name [Source]	FEMA No.	Regulation	Limitations
2-Propen-1-yl 3-phenyl-2-propenoate	See allyl cinnamate			
2-Propenyl propanoate	See allyl propionate			
Propenyl propyl disulfide		3227		
2-Propenyl sulfide	See allyl sulfide			
2-Propenyl 10-undecenoate	See allyl 10-undecenoate			
4-Propenyl veratrole	See isoeugenyl methyl ether			
Propionaldehyde	Propanol	2923	FDA §172.515	Synthetic flavor
Propionic acid	Propanoic acid	2924	FDA §182.1	GRAS, chemical preservative
Propionic acid, benzyl ester	See benzyl propionate			
Propionic acid, *p*-methoxybenzyl ester	See anisyl propionate			
Propionic acid, thioacrylic ester	See allyl thiopropionate			
Propionic acid, triglyceride	See glyceryl tripropanoate			
Propionylbenzene	See propiophenone			
Propiophenone		3469		
Propyl acetate	Acetic acid *n*-propyl ester	2925	FDA §172.515	Synthetic flavor
Propyl (iso) acetate	See isopropyl acetate			
Propylacetic acid	See valeric acid			
p-Propyl (iso) acetophenone	See *p*-isopropylacetophenone			
β-Propylacrolein	See 2-hexenal			
3-Propylacrylic acid	See *trans*-2-hexenoic acid			
β-Propylacrylic acid	See *trans*-2-hexenoic acid			
n-Propyl alcohol	1-Propanol	2928	FDA §172.515	Synthetic flavor
sec-Propyl alcohol	See isopropyl alcohol			
Propyl (iso) alcohol	See isopropyl alcohol			
Propyl aldehyde	See propionaldehyde			
3-Propylallyl alcohol	See 2-hexen-1-ol			
p-Propyl anisole	Dihydroanethole	2930	FDA §172.515	Synthetic flavor
p-Propyl (iso) benzaldehyde	See cuminaldehyde			

n-Propyl benzenecarboxylate	See propyl benzoate			
Propyl benzoate		2931	FDA §172.515	Synthetic flavor
Propyl (iso) benzoate	See isopropyl benzoate			
p-iso-Propylbenzyl alcohol	See *p*-isopropylbenzyl alcohol			
n-Propyl benzyl carbinol	See *α*-propylphenethyl alcohol			
Propyl butanoate	See propyl butyrate			
Propyl (iso) butanoate	See isopropyl butyrate			
Propyl butyrate		2934	FDA §172.515	Synthetic flavor
Propyl (iso) butyrate	See isopropyl butyrate			
γ-*n*-Propyl-*γ*-butyrolactone	See *γ*-heptalactone			
Propyl caproate	See propyl hexanoate			
Propyl (iso) caproate	See isopropyl hexanoate			
Propyl (iso) capronate	See isopropyl hexanoate			
Propyl carbinol	See butyl alcohol			
Propyl (iso) carbinol	See isobutyl alcohol			
Propyl chemosept	See propyl *p*-hydroxybenzoate			
Propyl cinnamate		2938	FDA §172.515	Synthetic flavor
Propyl (iso) cinnamate	See isopropyl cinnamate			
Propyl (iso) *o*-cresol	See carvacrol			
Propyl disulfide		3228	FDA §172.515	Synthetic flavor
Propyldithiopropane	See propyl disulfide			
2-Propylene acrolein	See *trans,trans*-2,4-hexadienal			
Propylene glycol	1,2-Propanediol	2940	FDA §182.1	GRAS, emulsifying agent and miscellaneous and/or general-purpose food additive
Propylene glycol-acetone ketal	See 2,2,4-trimethyl-1,3-oxacyclopentane			

REGULATORY STATUS OF DIRECT FOOD ADDITIVES (continued)

Material	Alternatate Name [Source]	FEMA No.	Regulation	Limitations
Propylene glycol alginate		2941	FDA §25.2 and §25.3	In French and salad dressings
			FDA §121.1015	1. Stabilizer in ice cream, frozen custard, ice milk, fruit sherbet, and water ices, up to 0.5%
				2. Emulsifier, stabilizer, or thickener in foods, except for standardized foods that do not provide for such use; product specifications apply
			FDA §121.1099	Defoaming agent
Propylene glycol alkaline			FDA §8.305	Extractant for annatto
Propylene glycol mono- and diesters of fats and fatty acids			FDA §121.1113	In foods; fats and fatty acids conforming with §121.1070
			MID	To emulsify shortenings sold in units not exceeding 6 lb or 1 gal fluid content
Propylene glycol monooctadecanoate	See propylene glycol stearate			
Propylene glycol monostearate	See propylene glycol stearate			
Propylene glycol octadecanoate	See propylene glycol stearate			
Propylene glycol stearate	Propylene glycol octadecanoate	2942		
Propyl ethanoate	See propyl acetate			
Propyl formate		2943	FDA §172.515	Synthetic flavor
Propyl (iso) formate	See isopropyl formate			
Propyl 2-furanacrylate	Propyl 3-(2-furyl) acrylate	2945	FDA §172.515	Synthetic flavor

Propyl 2-furoate		2946		
Propyl 3-(2-furyl) acrylate	See propyl 2-furanacrylate			
Propyl gallate	Tenox PG®	2947	FDA §182.1	GRAS, chemical preservative; max total antioxidants, 0.02% of fat or oil content, including essential (*volatile*) oil content of food
			FDA §121.1059	Antioxidant in chewing-gum base, alone or with butylated hydroxyanisole and/or butylated hydroxytoluene, up to 0.1% total antioxidant content
			MID	To retard rancidity in rendered animal fat or combination of such fat and vegetable fat, up to 0.01%, or in a combination of propyl gallate, BHA, BHT, glycine, and NDGA, up to 0.02%
			MID	Antioxidant to retard rancidity in frozen fresh pork sausages and freeze-dried meats, up to 0.01% based on fat content; can be used in combination with BHA, BHT, and/or NDGA, up to 0.02% in frozen fresh pork sausages and up to 0.01% in freeze-dried meats
Propyl heptanoate		2948	FDA §172.515	Synthetic flavor
Propyl heptoate	See propyl heptanoate			
Propyl heptylate	See propyl heptanoate			
Propyl hexanoate		2949	FDA §172.515	Synthetic flavor
Propyl (iso) hexanoate	See isopropyl hexanoate			

REGULATORY STATUS OF DIRECT FOOD ADDITIVES (continued)

Material	Alternatate Name [Source]	FEMA No.	Regulation	Limitations
Propyl (iso) hexylate	See isopropyl hexanoate			
Propyl (iso) hydrocinnamaldehyde	See 3-(*p*-isopropylphenyl) oropionaldehyde			
Propyl *p*-hydroxybenzoate	See propylparaben			
4-*n*-Propyl-4-hydroxybutanoic acid lactone	See γ-heptalactone			
5-Propyl-5-hydroxypentanoic acid lactone	See δ-octalactone			
Propylic alcohol	See propyl alcohol			
n-Propylidenebutyraldehyde	See 4-heptenal			
3-Propylidenephthalide		2952	FDA §172.515	Synthetic flavor and adjuvant
Propyl isobutyrate		2936	FDA §172.515	Synthetic flavor
Propyl (iso) isobutyrate	See isopropyl isobutyrate			
Propyl isopentanoate	See propyl isovalerate			
Propyl (iso) isopentanoate	See isopropyl isovalerate			
Propyl isovalerate		2960	FDA §172.515	Synthetic flavor
Propyl (iso) isovalerate	See isopropyl isovalerate			
Propyl isovalerianate	See propyl isovalerate			
Propyl mercaptan		—	FDA §172.515	Synthetic flavor
n-Propyl mercaptan	See propanethiol			
Propyl methanoate	See propyl formate			
Propyl (iso) methanoate	See isopropyl formate			
4-Propylmethoxybenzene	See *p*-propyl anisole			
2-Propyl (iso)-(3,5, or 6)-methoxypyrazine	See 2-methoxy-3(5 and 6)-isopropylpyrazine			
Propyl 3-methylbutanoate	See propyl isovalerate			
Propyl (iso) 3-methylbutanoate	See isopropyl isovalerate			

Propyl 3-methylbutyrate	See propyl isovalerate			
Propyl methyl carbinol	See 2-pentanol			
Propyl (iso) α-methylcrotonate	See isopropyl tiglate			
1-Propyl (iso)-4-methyl-2,4-cyclohexadiene	See α-phellandrene			
4-Propyl (iso)-1-methyl-1,5-cyclohexadiene	See α-phellandrene			
5-Propyl (iso)-2-methyl-1,3-cyclohexadiene	See α-phellandrene			
2-Propyl (iso)-5-methylcyclohexanol	See *d*-neo-menthol			
2-Propyl (iso)-5-methylcyclohexanone	See methone			
4-Propyl (iso)-1-methylcyclohexan-3-one	See menthone			
1-Propyl (iso)-4-methyl-2-cyclohexanone	See isopulegone			
4-Propyl (iso)-1-methyl-1-cyclohexen-3-one	See *d*-piperitone			
1-Propyl (iso)-4-methylcyclohex-2-yl acetate	See menthyl acetate			
1-Propyl (iso)-4-methylcyclohex-2-yl 3-butanote	See menthyl isovalerate			
Propyl methyl ketone	See 2-pentanone			
Propyl (iso) methyl ketone	See 4-methyl-2,3-pentanedione			
Propyl 2-methylpropanoate	See propyl isobutyrate			
Propyl (iso) 2-methylpropanoate	See isopropyl isobutyrate			
Propyl methyl trisulfide	See methyl propyl trisulfide			
Propylparaben	Propyl *p*-hydroxybenzoate	2951	FDA §182.1	GRAS, chemical preservative, up to 0.1%
			FDA §172.515	Synthetic Flavor
Propylparasept	See propyl paraben			
Propyl penethyl acetal	See acetaldehyde phenethyl propyl acetal			
α-Propylphenethyl alcohol	1-Phenyl-2-pentanol	2953	FDA §172.515	Synthetic flavor
2-Propylphenol	See *o*-propylphenol			

REGULATORY STATUS OF DIRECT FOOD ADDITIVES (continued)

Material	Alternatate Name [Source]	FEMA No.	Regulation	Limitations
o-Propylphenol		3522		
α-iso-Propyl phenylacetaldehyde	See 3-methyl-2-phenylbutyraldehyde			
p-Propyl (iso) phenylacetaldehyde	See *p*-isopropylphenylacetaldehyde			
Propyl phenylacetate		2955	FDA §172.515	Synthetic flavor
Propyl (iso) phenylacetate	See isopropyl phenylacetate			
n-Propyl *β*-phenylacrylate	See propyl cinnamate			
n-Propyl 3-phenylpropenoate	See propyl cinnamate			
3-(*p*-Propyl (iso)) phenylpropionaldehyde	See 3-(*p*-isopropylphenyl) propionaldehyde			
Propyl propanoate	See propyl propionate			
Propyl propionate		2958	FDA §172.515	Synthetic flavor
Propyl (iso) propionate	See isopropyl propionate			
Propyl thioacetate		3385		
Propylthiol	See propyl mercaptan			
Propyl (iso) tiglate	See isopropyl tiglate			
Propyl *α*-toluate	See propyl phenylacetate			
Propyl trisulfide	See dipropyl trisulfide			
Propyl valerianate		—		
δ-Propyl-*δ*-valerolactone	See *δ*-octalactone			
Protein, hydrolyzed	See hydrolyzed protein			
Protein, plant, hydrolyzed	See hydrolyzed plant protein			
Proteolytic enzymes (see also specific enzymes)	Includes *Aspergillus oryzae; A. flavusoryzae* group; promelin; ficin; papain		MID	With water, salt, and monosodium glutamate to soften tissues of beef cuts; tissue weight gain, 3% max
Protocatechualdehyde dimethyl ether	See veratraldehyde			
Protocatechualdehyde-3-methyl ether	See vanillin			
Protocatechuic aldehyde methylene ether	See piperonal			

Prunolide	See γ-nonalactone			
Pseudopinene	See β-pinene			
Psyllium seed, husk	[*Plantago psyllium* (L.); *P. arenaria* Waldst. and Kit.]		FDA §20	In frozen desserts
Pteroylglutamic acid	Folic acid		FDA §121.1134	In dietary supplements; max daily ingestion, 0.10 mg
Pulego (iso)	See isopulegol			
Pulegol (iso) acetate	See isopulegyl acetate			
Pulegone	*p*-Methyl-4(8)-en-3-one	2963	FDA §172.515	Synthetic flavor
Pyrazine ethanethiol	Pyrazinyl ethanethiol	3230		10 ppm max in finished food
Pyrazinemethanethiol	See 2-mercaptomethylpyrazine			
Pyrazinyl ethanethiol	See pyrazine ethanethiol			
Pyrazinyl methyl methyl sulfide	See pyrazinyl methyl sulfide			
Pyrazinyl methyl sulfide	Pyrazinyl methyl methyl sulfide	3231		10 ppm max in finished food
Pyridine		2966	FDA §172.515	Synthetic flavor
2-Pyridinemethanethiol	2-Pyridylmethanethiol	3232		
Pyridoxine hydrochloride	Vitamin B_6 hydrochloride		FDA §182.1	GRAS, nutrient and/or dietary supplement
2-Pyridylmethanethiol	See 2-pyridinemethanethiol			
Pyroacetic ether	See acetone			
Pyrocatechol monomethyl ether	See guaiacol			
Pyrogallol dimethyl ether	See 2,6-dimethoxyphenol			
Pyroguaiac acid	See guaiacol			
Pyroligneous acid	Pyroligneous vinegar	2967	FDA §3.201	Flavoring
Pyroligneous acid, extract		2968	FDA §172.515	Synthetic flavor
Pyromucic aldehyde	See furfural			
Pyromucyl acetone	See 1-(2-furyl)-1,3-butanedione			
Pyroracemic acid	See pyruvic acid			
Pyrrole		3386		
Pyrrolidine		3523		
2-Pyrrolidinecarboxylic acid	See L-proline			
Pyruvaldehyde	Pyruvic aldehyde	2969	FDA §172.515	Synthetic flavor
Pyruvic acid	Pyroracemic acid	2970	FDA §172.515	Synthetic flavor
Pyruvic aldehyde	See pyruvaldehyde			

REGULATORY STATUS OF DIRECT FOOD ADDITIVES (continued)

Material	Alternatate Name [Source]	FEMA No.	Regulation	Limitations
Quack grass	See dog grass, extract [*Agropyron repens* (L.) Beauv.]			
Quadrafos®	See sodium hexametaphosphate			
Quassia, extract	[*Picrasma excelsa* (Sw.) Planch.; *Quassia amara* (L.)]	2971	FDA §712.510	Natural flavor
Quebracho bark, extract	[*Aspidosperma quebrachoblanco* Schlecht; *Schinopsis lorentzii* (Griseb.) Engl.; or *Quebrachia lorentzii* Griseb.]	2972	FDA §172.510	Natural flavoring substances and natural substances used in conjunction with flavors
Quick grass	See dog grass, extract [*Agropyron repens* (L.) Beauv.]			
Quilleia	Soapbark [*Quillaia saponaria* Molina]	2973	FDA §172.510	Natural flavor
Quince seed, extract	[*Cydonia oblonga* Miller]	2974	FDA §182.1	GRAS, solvent-free natural extractive used in conjunction with natural flavoring
Quinine	[*Cinchona officinalis* (L.) (*C. ledgeriana* Moens)]		FDA §121.1081	Hydrochloride or sulfate of quinine, as flavor in carbonated beverages, up to 83 ppm as quinine
Quinine bisulfate		2975		
Quinine chloride	See quinine hydrochloride			
Quinine hydrochloride	Quinine chloride	2976		
Quinine sulfate		2977		
Quinoline		3470		
Quinoline (iso)	See isoquinoline			
Racemic acid	See tartaric acid			
Raldeine	See methyl α-ionone			
Raldeine gamma	See α-iso-methylionone			
Raspberry ketone methyl ether	See 4-(*p*-methoxyphenyl)-2-butanone			

Red saunders	Red sandalwood [*Pterocarpus santalinus* (L.) f.]		FDA §172.510	Natural flavor in alcoholic beverages only
Rennin	Rennet		FDA §182.1	GRAS, miscellaneous and/or general-purpose food additive
Resin guaiac			FDA	0.1% max of fat content of food
			MID	To retard rancidity in rendered animal fat or combination of such fat and vegetable fat, up to 0.10%
Resin, natural			FDA §121.1059	Terpene resin in chewing-gum base, consisting of polymers of α-pinene; product specifications apply
Resorcinol	1,3-Benzenediol	3589		See *Food Technol.*, February 1978
	m-Dihydroxybenzene	3589		See *Food Technol.*, February 1978
Resorcinol dimethyl ether	See *m*-dimethoxybenzene			
Rhatany, extract	[*Krameria triandra* Ruiz and Pav.; *K. argentea* Mart.]	2979	FDA §172.510	Natural flavor
Rhizopus oryzae carbohydrase	[*Rhizopus oryzae*]		FDA §121.1165	In production of dextrose from starch; product specifications apply
Rhodinal	See citronellal			
Rhodinol	*l*-Citronellol; 3,7-dimethyl-7-octen-1-ol	2980	FDA §172.515	Synthetic flavor
Rhodinolic acid	See 3,7-dimethyl-6-octenoic acid			
Rhodinyl acetate		2981	FDA §172.515	Synthetic flavor
Rhodinyl butyrate		2982	FDA §172.515	Synthetic flavor
Rhodinyl ethanoate	See rhodinyl acetate			
Rhodinyl formate		2984	FDA §172.515	Synthetic flavor
Rhodinyl isobutyrate		2983	FDA §172.515	Synthetic flavor
Rhodinyl isopentanoate	See rhodinyl isovalerate			
Rhodinyl isovalerate		2987	FDA §172.515	Synthetic flavor
Rhodinyl isovalerianate	See rhodinyl isovalerate			
Rhodinyl methanoate	See rhodinyl formate			
Rhodinyl 3-methylbutanoate	See rhodinyl isovalerate			
Rhodinyl 2-methylpropanoate	See rhodinyl isobutyrate			
Rhodinyl phenylacetate		2985	FDA §172.515	Synthetic flavor
Rhodinyl propanoate	See rhodinyl propionate			
Rhodinyl propionate		2986	FDA §172.515	Synthetic flavor

REGULATORY STATUS OF DIRECT FOOD ADDITIVES (continued)

Material	Alternatate Name [Source]	FEMA No.	Regulation	Limitations
Rhodinyl α-toluate	See rhodinyl phenylacetate			
Rhubarb, garden root	[*Rheum rhaponticum* (L.)]	—	FDA §172.510	Natural flavor in alcoholic beverages only
Rhubarb, root	[*R. officinale* Baill.; *R. palmatum* (L.); or other *Rheum* spp. (except *R. rhaponticum* (L.)) or hybrids of *Rheum* grown in China]	—	FDA §172.510	Natural flavor
Riboflavin	Vitamin B_2		FDA §182.1	GRAS, nutrient and/or dietary supplement
			FDA §8.323	Coloring foods generally; not to be used in standardized foods unless standards provide for use of added color; specifications in *Food Chemicals Codex* apply
Riboflavin-5-phosphate			FDA §182.1	GRAS, nutrient and/or dietary supplement
Rice, bran, wax			FDA §121.1098	50 ppm max in candy; coating, 50 ppm max in fresh fruits and fresh vegetables; coating, 2.5% max in chewing gum; plasticizing material; product specifications apply
			FDA §121.1059	Plasticizer (softener) in chewing-gum base
Ricinus, oil	See castor, [*Ricinus communis* (L.)]			
Rosal	See phenylacetaldehyde dimethyl acetal			
Rose	Otto of rose, attar of rose [*Rosa alba* (L.); *R. centifolia* (L.); *R. damascena* Mill.; *R. gallica* (L.); and varieties of these species]		FDA §182.1	GRAS, natural flavor extractive
Rose, absolute	[*R. alba* (L.); *R. centifolia* (L.); *R. damascena* Mill.; *R. gallica* (L.); and varieties of these species	2988	FDA §182.1	GRAS, natural flavor extractive

Rose, buds, flowers, fruits (hips), and leaves	[*Rosa* spp.]		FDA §182.1	GRAS, natural flavor extractive
Rose, Bulgarian, true Otto, oil	[*R. damascena* Mill.]	2989	FDA §182.1	Natural flavor
Rose geranium	See geranium, rose			
Rose hips, extract	[*Rosa* spp.]	2990	FDA §182.1	Natural flavor
Roselle	[*Hibiscus sabdariffa* (L.)]		FDA §172.510	Natural flavor in alcoholic beverages only
Rosemary	[*Rosmarinus officinalis* (L.)]	2991	FDA §182.1	GRAS, natural flavor and natural flavor extractive
Rosemary, oil	[*Rosmarinus officinalis* (L.)]	2992	FDA §182.1	Natural flavor
Rose oxide	See tetrahydro-4-methyl-2-(2-methylpropen-1-yl) pyran			
Rose water, stronger	[*Rosa centifolia* (L.)]	2993		
Rosidinha	Rosadinha [*Micropholis* (also known as *Sideroxylon*) spp.]		FDA §121.1059	Natural (coagulated or concentrated latex) masticatory substance in chewing-gum base
Rosin	Colophony; abietic anhydride [*Pinus palustris* Mill. and other *Pinus* spp.]	—	FDA §172.510	Natural flavor in alcoholic beverages only
Rosin and rosin derivatives			FDA §8.390	Diluent in inks for marking fruits and vegetables; inks for marking gum, confectionery, and food supplements in tablet form; and mixtures for coloring eggshells (no penetration through eggshell into egg); specifications in §121.1059 apply
Ruby wood	See sandalwood, red [*Pterocarpus santalinus* (L.) f.]			
Rue	[*Ruta graveolens* (L.)]	2994	FDA §182.1	GRAS, natural flavor and natural flavor extractive
Rue, oil	[*R. graveolens* (L.)]	2995	FDA §182.1	Natural flavor
Rum ether	Ethyl oxyhydrate	2996	FDA §172.515	Synthetic flavor
Sabinenehydrate	See 4-thujanol			
Saccharin	2,3-Dihydro-3-oxobenzisosulfonazole; saccharin, insoluble		FDA §182.1	GRAS, nonnutritive sweetener
			MID	To sweeten bacon, at 0.01%
Saccharin, insoluble	See saccharin			
Saccharin, sodium salt	Saccharin, soluble	2997	FDA §182.1	GRAS, in dietary supplements

REGULATORY STATUS OF DIRECT FOOD ADDITIVES (continued)

Material	Alternatate Name [Source]	FEMA No.	Regulation	Limitations
Saccharin, soluble	See saccharin, sodium salt			
Saffron	[*Crocus sativus*(L.)]	2998	FDA §182.1	GRAS, natural flavor and natural flavor extractive
			FDA §8.11	In foods, except in standardized foods where standards do not authorize its use
			MID	To color sausage casings, oleomargarine, or shortening, and in marking or branding ink for products; may be mixed with approved synthetic dyes or harmless inert material, such as common salt or sugar
Saffron, extract	[*C. sativus*(L.)]	2999	FDA §182.1	Natural flavor
Safranal	See 2,6,6-trimethylcyclohexa-1,3-dienyl methanal			
Sage	[*Salvia officinalis*(L.)]	3000	FDA §182.1	GRAS, natural flavor and natural flavor extractive
Sage, Greek	[*S. triloba*(L.)]		FDA §182.1	GRAS, natural flavor and natural flavor extractive
Sage, Mexican	See oregano			
Sage, oil	[*S. officinalis*(L.)]	3001	FDA §182.1	Natural flavor
Sage, oleoresin	[*S. officinalis*(L.)]	3002	FDA §182.1	Natural flavor
Sage, Spanish	[*S. lavendulaefolia* Vahl.]	3003	FDA §182.1	GRAS, natural flavor extractive
Saigon cinnamon	See cinnamon bark, Saigon, extract [*Cinnamomum loureirii* Nees] or cinnamon leaf, oil [*Cinnamomum* spp.]			
St. John's bread	See carob bean extract [*Ceratonia siliqua*(L.)]			
St. Johnswort leaves, flowers, and caulis	[*Hypericum perforatum*(L.)]	—	FDA §172.510	
Sal ethyl	See ethyl salicylate			
Salicylal	See salicylaldehyde			

Salicylaldehyde	*o*-Hydroxybenzaldehyde	3004	FDA §172.515	Synthetic flavor
Salicylic acid, benzyl ester	See benzyl salicylate			
Salicylic acid, ethyl ester	See ethyl salicylate			
Salicylic acid, isopentyl ether	See isoamyl salicylate			
Salicylic ether	See ethyl salicylate			
Salts of fatty acids			FDA §121.1071	Binder, emulsifier, and anticaking agents; product specifications apply
			FDA §8.201	Diluent in aqueous suspension of Citrus Red No. 2; conforming with §121.1071
Sandalwood, East Indian	See sandalwood, yellow [*Santalum album* (L.)]			
Sandalwood, East Indian, oil	See sandalwood, yellow, oil [*S. album* (L.)]			
Sandalwood, red	[*Pterocarpus santalinus* (L.) f.]	—	FDA §172.510	
Sandalwood, West Indian	See amyris [*Amyris balsamifera* (L.)]			
Sandalwood, West Indian, oil	See amyris, oil [*A. balsamitera* (L.)]			
Sandalwood, white	See sandalwood, yellow [*S. album* (L.)]			
Sandalwood, white, oil	See sandalwood, yellow, oil [*S. album* (L).]			
Sandalwood, yellow	[*S. album* (L.)]	—	FDA §172.510	
Sandalwood, yellow, oil	Yellow or East Indian sandalwood [*S. album* (L.)]	3005	FDA §172.510	Natural flavoring substances and natural substances used in conjunction with flavors
Sandarac	[*Tetraclinis articulata* (Vahl.) Mast.]	—	FDA §172.510	Natural flavor in alcoholic beverages only
Sanderswood, red	See sandalwood, red [*P. santalinus* (L.) f.]			
Santalol, α and β	Argeol	3006	FDA §172.515	Synthetic flavor
Santalyl acetate		3007	FDA §172.515	Synthetic flavor
Santalyl phenylacetate		3008	FDA §172.515	Synthetic flavor
α-Santalyl phenylacetate	See santalyl phenylacetate			
β-Santalyl phenylacetate	See santalyl phenylacetate			
Santalyl α-toluate	See santalyl phenylacetate			
Saponin	See quillaia [*Quillaa saponaria* Molina]			

REGULATORY STATUS OF DIRECT FOOD ADDITIVES (continued)

Material	Alternatate Name [Source]	FEMA No.	Regulation	Limitations
Sarsaparilla, extract	[*Smilax aristolochiaefolia* Mill. (Mexican sarsaparilla); *S. regelii* Killip and Morton (Honduras sarsaparilla); *S. febrifuga* Kunth (Ecuadorean sarsaparilla); or undetermined species of *Smilax* (Ecuadorean or Central American sarsaparilla)]	3009	FDA §172.510	Natural flavor
Sassafras, bark, extract	[*Sassafras albidum* (Nuttall) Nees (Fam. Lauraceae)]	3010	FDA §121.1097	Safrole-free extract as flavoring in food; product specifications apply
Sassafras, leaf	[*Sassafras albidum* (Nuttall) Nees (Fam. Lauraceae)]	3011	FDA §172.510	Natural flavor; safrole-free
Saunders, red	See sandalwood, red [*Pterocarpus santalinus* (L.) f.]			
Saunders, white	See sandalwood, yellow [*Santalum album* (L.)]			
Saunders, white, oil	See sandalwood, yellow, oil [*Santalum album* (L.)]			
Savory, summer	[*Satureia hortensis* (L.) Satureia]	3012	FDA §182.1	GRAS, natural flavor and natural flavor extractive
Savory, summer, oil	[*Satureia hortensis* (L.) Satureia]	3013	FDA §182.1	Natural flavor
Savory, summer, oleoresin	[*Satureia hortensis* (L.) Satureia]	3014	FDA §182.1	Natural flavor
Savory, winter	[*Satureia montana* (L.) Satureia]	3015	FDA§182.1	GRAS, natural flavor and natural flavor extractive
Savory, winter, oil	[*Satureia montana* (L.) Satureia]	3016	FDA §182.1	Natural flavor
Savory, winter, oleoresin	[*Satureia montana* (L.) Satureia]	3017	FDA §182.1	Natural flavor
Schinus molle, oil	[*Schinus molle* (L.)]	3018	FDA §182.1	GRAS, natural flavor extractive
Senecioic acid	See 3-methylcrotonic acid			
Senna, Alexandria	[*Senna acutifolia* Nect., also called *Cassia acutifolia* Delile]	—	FDA §182.1	Natural flavoring substances and natural substances used in conjunction with flavors
Septos	See methyl *p*-hydroxybenzoate			
Serine (L- and DL-forms)			FDA §182.1	GRAS, nutrient and/or dietary supple-

				ment, as free acid, hydrochloride salt, hydrate, or anhydrous form
Serpentaria	Virginia snakeroot [*Aristolochia serpentaria* (L.)]	—	FDA §172.510	In alcoholic beverages only
Sesame	[*Sesamum indicum* (L.)]		FDA §182.1	GRAS, natural flavor
Shellac, purified			FDA §8.309 and §8.6200	Diluent in inks for marking gum, confectionery, and food supplements in tablet form; inks for branding pharmaceutical forms; food-grade
Silicon dioxide	Silica		FDA §121.229	Anticaking agent and/or grinding aid in feed and feed components: 1. 2% max in BHT 2. 1% max in methionine hydroxy analogue and its calcium salts 3. 0.8% max in piperazine or piperazine salts 4. 1% max in sodium propionate 5. 1% max in urea 6. 3% max in vitamins Product specifications apply
			FDA §121.1058	1. Anticaking agent, 2% max; not to be used in infant foods, except when present as anticaking agent at 2% max in salt and salt substitutes used as components of these foods 2. Stabilizer in production of beer, to be removed from the beer by filtration prior to final processing 3. Adsorbent for DL-α-tocopheryl acetate and pantothenol in foods for special dietary use Product specifications apply
			FDA §8.390	Diluent in inks for marking fruits and vegetables, up to 2% of the ink solids; specifications in §121.1058 apply
Simaruba bark	[*Simaruba amara* Aubl.]	—	FDA §172.510	Natural flavor in alcoholic beverages only

REGULATORY STATUS OF DIRECT FOOD ADDITIVES (continued)

Material	Alternatate Name [Source]	FEMA No.	Regulation	Limitations
α-Sinensal	See 2,6-dimethyl-10-methylene-2,6,11-dodecatrienal			
Sionon®	See *d*-sorbitol			
Skatole	3-Methylindole	3019	FDA §172.515	Synthetic flavor
Sloe berries	Blackthorn [*Prunus spinosa* (L.)]	3020	FDA §182.1	GRAS, natural flavor extractive
Sloe berries, extract	[*P. spinosa* (L.)]	3021	FDA §182.1	Natural flavor
Sloe berries, extract solid	[*P. spinosa* (L.)]	3022	FDA §182.1	Natural flavor
Smallage	See lovage [*Levisticum officinale* Koch]			
Smallage, extract	See lovage, extract [*L. officinale* Koch]			
Smallage, oil	See lovage, oil [*L. officinale* Koch]			
Smellage	See lovage [*L. officinale* Koch]			
Smellage, extract	See lovage, extract [*L. officinale* Koch]			
Smellage, oil	See lovage, oil [*L. officinale* Koch]			
Smoke flavoring			MID	To flavor various meat products
Snakeroot, Canadian, oil	Wild ginger [*Asarum canadense* (L.)]	3023	FDA §172.510	Natural flavor
Snakeroot, Virginia	See serpentaria			
Soap bark	See quillaia [*Quillaia saponaria* Molina]			
Sodium acetate		3024	FDA §182.1	GRAS, miscellaneous and/or general-purpose food additive
Sodium acid phosphate			FDA §182.1	GRAS, sequestrant
Sodium acid pyrophosphate	Disodium dihydrogen pyrophosphate		FDA §182.1	GRAS, miscellaneous and/or general-purpose food additive
			MID	To decrease amount of cooked-out juices in cured hams, pork shoulder picnics, and loins, canned hams and pork shoulder picnics, chopped ham, and bacon; 5.0% of phosphate in pickle at 10% pump level; 0.5% of phos-

				phate in product (only clear solution may be injected into product)
Sodium alginate	See alginates — sodium, calcium, and ammonium salts			
Sodium aluminosilicate	sodium silicoaluminate		FDA §182.1	GRAS, anticaking agent, up to 2%
Sodium aluminum phosphate			FDA §182.1	GRAS, miscellaneous and/or general-purpose food additive
Sodium aluminum sulfate			FDA §15	Bleaching agent in flour, at not more than 6 parts by weight, alone or with potassium alum, calcium sulfate, magnesium carbonate, dicalcium phosphate, tricalcium phosphate, starch, and/or calcium carbonate, with 1 part by weight benzoyl peroxide
Sodium ascorbate	Ascorbic acid sodium derivative		FDA §182.1	GRAS, chemical preservative
			MID	To accelerate color fixing in cured pork and beef cuts and in cured, comminuted meat products; 87.5 oz/100 gal pickle at 10% pump level; 7/8 oz/100 lb of meat or meat byproduct; no significant amount of moisture shall be added to product
Sodium benzoate		3025	FDA §182.1	GRAS, chemical preservative, up to 0.1%
			MID	To retard flavor reversion of oleomargarine, at 0.1%
Sodium bicarbonate	Sodium hydrogen carbonate		FDA §182.1	GRAS, miscellaneous and/or general-purpose food additive
			MID	To separate fatty acids and glycerol in rendered fats; must be eliminated during manufacturing process
			MID	To neutralize excess acidity and clean vegetables in rendered fats, soups, and curing pickle
Sodium bisulfite	Sodium acid sulfite		FDA §182.1	GRAS, chemical preservative; not for use in meats or in foods recognized as source of vitamin B_1

REGULATORY STATUS OF DIRECT FOOD ADDITIVES (continued)

Material	Alternatate Name [Source]	FEMA No.	Regulation	Limitations
Sodium *m*-bisulfite			FDA §182.1	GRAS, chemical preservative; not for use in meats or in foods recognized as source of vitamin B_1
Sodium calcium aluminosilicate, hydrated	Sodium calcium silicoaluminate		FDA §182.1	GRAS, anticaking agent
Sodium carbonate			FDA §182.1	GRAS, miscellaneous and/or general-purpose food additive
			FDA §8.303	In caramel (including bicarbonate)
			MID	To refine rendered fats; must be eliminated during manufacturing process
Sodium carbonate with sodium mono- and dimethyl naphthalene sulfonates			FDA §121.1198	In potable water system to reduce hardness and aid in sedimentation and coagulations by raising the pH for efficient utilization of other coagulation materials; sodium mono- and dimethyl sulfonates are used in crystallization of sodium carbonate at 250 ppm max of sodium carbonate; product specifications apply
Sodium carboxymethylcellulose	Carboxymethylcellulose; cellulose gum; CMC	2239	FDA §182.1	GRAS, miscellaneous and/or general-purpose food additive; product specifications apply
			FDA §19	In specified cheese products, up to 0.8% by weight
			MID	To extend and stabilize baked pie
Sodium caseinate			FDA §182.1	GRAS, miscellaneous and/or general-purpose food additive
			MID	To bind and extend imitation sausage, nonspecific loaves, soups, and stews
Sodium chloride	Table salt		FDA-GRAS	
Sodium chlorite			FDA §121.1031	Modifier for food starch, up to 0.5%
Sodium citrate	Citratin	3026	FDA §18.520	In evaporated milk, alone or with diso-

				dium phosphate, up to 0.1% by weight of finished product
			FDA §182.1	GRAS, sequestrant and miscellaneous and/or general-purpose food additive
			MID	To prevent clotting of fresh beef blood at 0.2% with or without water; product specifications apply
			MID	To accelerate color fixing in cured pork and beef cuts and in cured, comminuted meat food products; in cured products to replace up to 50% of ascorbic acid, erythorbic acid, sodium ascorbate, or sodium erythorbate used
Sodium dehydroacetate			FDA §121.1089	Preservative for cut or peeled squash; 65 ppm max (expressed as dehydroacetic acid) remaining in or on the prepared squash; product specifications apply
Sodium diacetate	Sodium acid acetate		FDA §17 FDA §182.1 FDA-GRAS	In bakery products, up to 0.4% of flour GRAS, sequestrant
Sodium dihydrogen phosphate				
Sodium dimethyl naphthalene sulfonates			FDA §121.1198	In crystallization of sodium carbonate, 250 ppm max of sodium carbonate; sodium carbonate, so produced, used in potable water systems; product specifications apply
Sodium erythorbate			MID	To accelerate color fixing in cured pork and beef cuts and in cured, comminuted meat food products; 87.5 oz/100 gal pickle at 10% pump level; 7/8 oz/100 lb of meat or meat byproduct; no significant amount of moisture added to product
Sodium gluconate	Gluconic acid sodium salt		FDA §182.1	GRAS, sequestrant
Sodium hexametaphosphate		3027	FDA §182.1	GRAS, sequestrant
			MID	To decrease amount of cooked-out juices in cured hams, pork shoulder

REGULATORY STATUS OF DIRECT FOOD ADDITIVES (continued)

Material	Alternatate Name [Source]	FEMA No.	Regulation	Limitations
				picnics, and loins, canned hams and pork shoulder picnics, chopped ham, and bacon; 5.0% of phosphate in pickles at 10% pump level; 0.5% of phosphate in product (only clear solution may be injected into product)
Sodium hydroxide	Caustic soda		FDA §182.1	GRAS, Miscellaneous and/or general-purpose food additive
			FDA §121.1031	Modifier for food starch, up to 1.0%
			FDA §8.303	In production of caramel
Sodium lauryl sulfate	Dodecyl sodium sulfate		FDA §121.1012	1. Emulsifier in or with egg whites — 1000 ppm max in egg white solids, 125 ppm max in frozen egg whites, 125 ppm max in liquid egg whites 2. Whipping agent, 5000 ppm max by weight of gelatin used in preparation of marshmallows Product specifications apply
Sodium metaphosphate (see also sodium hexametaphosphate)			FDA §182.1	GRAS, sequestrant
			FDA §19	Emulsifier, up to 3% by weight of specified cheeses
Sodium mono- and dimethyl naphthalene sulfonates			FDA §121.1198	In crystallization of sodium carbonate, 250 ppm max; sodium carbonate, so produced, used in potable water systems; product specifications apply
Sodium nitrate	Chile saltpeter		FDA §121.1063	1. Preservative and color fixative, with or without sodium nitrite; 500 ppm max in smoked and cured sablefish, salmon, and shad; sodium nitrite in finished product, 200 ppm max

			2. In meat-curing preparations for home-curing of meat and meat products (including poultry and wild game); directions must limit amount of sodium nitrate to 500 ppm max and sodium nitrite to 200 ppm max in finished meat product; product specifications apply
		MID	Source of nitrite in cured products; 7 lb/100 gal pickle; 3.5 oz/100 lb meat (dry cure); 2.75 oz/10 lb chopped meat
Sodium nitrite	Erinitrit	FDA §121.223	20 ppm max alone as preservative and color fixative in canned pet food containing fish, meat, and fish and meat byproducts
		FDA §121.1064	1. Preservative and color fixative: 10 ppm max in smoked, cured tunafish products; 200 ppm max in smoked cured sablefish, salmon, and shad; may be used with sodium nitrate at 500 ppm max 2. In meat-curing preparations for home-curing of meat and meat products (including poultry and wild game) with sodium nitrate; directions must limit sodium nitrate to 500 ppm max and sodium nitrite to 200 ppm max Product specifications apply
		MID	To fix color in cured products; 2 lb/100 gal pickle at 10% pump level; 1 oz/100 lb meat (dry cure); ¼ oz/100 lb chopped meat and/or meat byproduct; 200 ppm max of nitrite in finished product; product specifications apply
Sodium pantothenate		FDA §182.1	GRAS, nutrient and/or dietary supplement
Sodium pectinate		FDA §182.1	GRAS, miscellaneous and/or general-purpose food additive

REGULATORY STATUS OF DIRECT FOOD ADDITIVES (continued)

Material	Alternatate Name [Source]	FEMA No.	Regulation	Limitations
Sodium phosphate, dibasic	See disodium phosphate			
Sodium phosphate (di- and monobasic)			FDA §8.303	In production of caramel
Sodium phosphate (di-mono-, and tribasic)			FDA §182.1	GRAS, nutrient and/or dietary supplement
			FDA §182.1	GRAS, sequestrant and miscellaneous and/or general-purpose food addtivie
Sodium potassium tartrate			FDA §182.1	GRAS, sequestrant and miscellaneous and/or general-purpose food additive
Sodium propionate	Propionic acid sodium salt		FDA §17	In bakery products, alone or with calcium propionate, up to 0.32% by weight of flour in bread, enriched bread, milk bread, and raisin bread, and up to 0.38% of flour used in whole-wheat bread
			FDA	In bread, up to 0.32% by weight of flour
			FDA §182.1	GRAS, chemical preservative
			MID	Alone or with calcium propionate, to retard mold growth in pizza crust, up to 0.32% by weight of flour; product specifications apply
Sodium pyrophosphate			FDA §182.1	GRAS, sequestrant
			MID	To decrease amount of cooked-out juices cured hams, pork shoulder picnics, and loins, canned hams, and pork shoulder picnics, chopped ham, and bacon; 5.0% of phosphate in pickle at 10% pump level; 0.5% of phosphate in product (only clear solution may be injected into product)
Sodium saccharin			FDA §182.1	GRAS, nonnutritive sweetener

Sodium salts of fatty acids		FDA §121.1071	Binder, emulsifier and anticaking agent in food; product specifications apply
Sodium sesquicarbonate		FDA §182.1	GRAS, miscellaneous and/or general-purpose food additive
Sodium silicoaluminate, precipitated, hydrated		FDA	2% max in salt; 2% max in seasoning; 1% max in sugar; 5% max in baking powder
Sodium sorbate		FDA §182.1	GRAS, chemical preservative
Sodium stearate		FDA §172.860 and §172.863	Cleared in food with GMP as a binder, emulsifier, and anticaking agent
		FDA §172.615	Cleared as plasticizer in chewing-gum base
Sodium stearoyl-2-lactylate		FDA §172.846	Cleared for use as emulsifier, dough conditioner, or whipping agent, where standards do not apply in icings, fillings, toppings, puddings, baked goods, pancakes, waffles, and prepared mixes of these and in liquid and solid milk or cream substitutes in coffee beverages; product standards apply
Sodium stearyl fumarate		FDA §121.1183	1. Dough conditioner in yeast-leavened products, 0.5% max by weight of flour 2. Conditioning agent in dehydrated potatoes, 1% max; product specifications apply
Sodium succinate	See disodium succinate		
Sodium sulfate		FDA §121.1059	Miscellaneous food additive in chewing-gum base
		FDA §8.303	In production of caramel
Sodium sulfide	Sodium monosulfide	FDA §121.1059	Miscellaneous food additive in chewing-gum base, for use as reaction-control agent in synthetic polymer production
Sodium sulfite		FDA	200 to 300 ppm in molases, dried fruits, and foods that are not good sources of vitamin B_1
		FDA §182.1	GRAS, chemical preservative; not for use in meats or in foods recognized as source of vitamin B_1

REGULATORY STATUS OF DIRECT FOOD ADDITIVES (continued)

Material	Alternatate Name [Source]	FEMA No.	Regulation	Limitations
			FDA §8.303	In production of caramel
Sodium sulfoacetate derivative of mono- and diglycerides			MID	To flavor various meat products, at 0.5%
Sodium tartrate			FDA §19	Emulsifier, up to 3% by weight of specified cheeses
			FDA §182.1	GRAS, sequestrant
Sodium tetrapyrophosphate			FDA §182.1	GRAS, sequestrant
Sodium thiosulfate	Sodium hyposulfite		FDA §182.1	GRAS, sequestrant in salt, up to 0.1%
Sodium trimetaphosphate			FDA §121.1031	In food starch — modified, alone, or as part of successive treatment with sodium tripolyphosphate and sodium trimetaphosphate; residual phosphate, 0.4% max calculated as phosphorus
Sodium tripolyphosphate	Sodium triphosphate		FDA §182.1	GRAS, sequestrant and miscellaneous and/or general-purpose food additive
			FDA §121.1031	In food starch-modified, for successive treatment of starch with sodium tripolyphosphate and sodium trimetaphosphate; residual phosphate, 0.4% max calculated as phosphorus
			MID	To decrease amount of cooked-out juices in cured hams, pork shoulder picnics, and loins, canned hams and pork shoulder picnics, chopped ham, and bacon; 5.0% of phosphate in pickle at 10% pump level; 0.5% of phosphate in product (only clear solution may be injected into product)
			FDA §8.201	Diluent for aqueous suspension of Citrus Red No. 2, up to 0.05%
Sorbic acid	2,4-Hexadienoic acid		FDA §182.1	GRAS, chemical preservative
Sorbic aldehyde	See *trans, trans*-2,4-hexadienal			

Sorbitan monostearate	3028	FDA §121.1029	Emulsifier, alone or with polyoxyethylene (20) sorbitan monostearate: 1. In whipped vegetable oil topping, alone or with polysorbate 60, polysorbate 80, and/or polyoxyethylene (20) sorbitan tristearate; total additive(s), 0.40% max 2. In cakes and cake mixes, 0.61% max alone; 0.61% max in combination with polyoxyethylene (20) sorbitan tristearate (0.32% max) and/or polysorbate 60 (0.46% max); total additives, 0.66% max 3. Alone in nonstandardized confectionery coatings, 10,000 ppm max; 10,000 ppm max in combination with polyoxyethylene (20) sorbitan monostearate (5000 ppm max); total additive(s), 10,000 ppm max 4. In cake icings and cake fillings, 0.70% max alone; 0.70% max in combination with polyoxyethylene (20) sorbitan tristearate (0.32% max) and/or polysorbate 60 (0.46% max); total additives, 1.0% max 5. In solid-state, vegetable-fat-water emulsions for use in beverage coffee as substitutes for milk or cream; 0.4% max, alone or with polyoxyethylene (20) sorbitan tristearate and/or polysorbate 60 6. Rehydration aid in production of active dry yeast, alone at 1% max by weight of dry yeast Product specifications apply
		FDA §121.1099	Defoaming agent, conforming with §121.1029
		FDA §172.515	Synthetic flavor

REGULATORY STATUS OF DIRECT FOOD ADDITIVES (continued)

Material	Alternatate Name [Source]	FEMA No.	Regulation	Limitations
			MID	To emulsify shortenings sold in units not exceeding 6 lb or 1 gal fluid content; alone at 1% or with polysorbate 60 or polysorbate 80; total additives, 1% max
Sorbitol	D-Glucitol; Sorbit®	3029	FDA §182.1	GRAS, nutrient and/or dietary supplement in foods for special dietary use, up to 7%
			FDA §121.1053	Food additive
D-Sorbitol	See sorbitol			
Soy Protein, isolated	See isolated soy protein			
Span® 60	See sorbitan monostearate			
Spanish hops	See dittany of Crete [*Origanum dictamnus* (L.)]			
Spanish "origanum"	See *Thymus capitatus* Hoff and Link			
Spearmint	[*Mentha spicata* Houds (L.)]	3030	FDA §182.1	GRAS, natural flavor and natural flavor extractive
Spearmint, extract	[*M. spicata* Houds (L.)]	3031	FDA §182.1	Natural flavor
Spearmint, oil	[*M. spicata* Houds (L.)]	3032	FDA §182.1	Natural flavor
Spike lavender	See Lavender, spike, oil [*Lavandula latifolia* Vill. (*L. spica* D.C.)]			
Spikenard extract		—		
Spiral flag, oil	See costus root, oil [*Saussurea lappa* Clarke]			
Spirit of nitrous ether	See ethyl nitrite			
Spruce, oil	[*Tsuga* and *Picea* spp.]	3034	FDA §172.510	
Spruce, needles and twigs	[*Picea glauca* (Moench) Voss; or *P. mariana* (Mill.) BSP]		FDA §172.510	Natural flavor
Stannous chloride	Tin chloride		FDA §182.1	GRAS, chemical preservative, up to 0.0015% calculated as tin
Star anise	See anise, star			
Starch hydrolysates			FDA §8.303	In production of caramel
Starter distillate			MID	To flavor oleomargarine

Stearic acid	Octadecanoic acid	3035	FDA §121.1059	Plasticizing material (softener) in chewing-gum base, complying with §121.1070
(mono)Stearin	See glyceryl monostearate			
Stearyl citrate			FDA §182.1	GRAS, sequestrant, up to 0.15%
			MID	To protect flavor of oleomargarine, at 0.15%
Stearyl-2-lactylic acid			MID	To emulsify shortening to be used for cake icings and fillings, at 3%
Stearyl monoglyceridyl citrate			FDA §121.1080	Emulsion stabilizer in or with shortenings containing emulsifiers; product specifications apply
			MID	To emulsify shortenings
Sterculia	See karaya, gum [*Sterculia urens* Roxb.]			
Storax	Styrax [*Liquidambar orientalis* Mill. or *L. styraciflua* (L.)]	3036	FDA §172.510	Natural flavor
Storax, extract	See styrax, extract [*Liquidambar* spp.]			
Strawberry aldehyde (see also ethyl methyl phenylglycidate)	[*Fragaria vesca* (L.)]		FDA §182.1	GRAS, synthetic flavor
Styralyl acetate	See α-methylbenzyl acetate			
Styralyl alcohol	See α-methylbenzyl alcohol			
Styralyl butyrate	See α-methylbenzyl butyrate			
Styralyl formate	See α-methylbenzyl formate			
Styralyl isobutyrate	See α-methylbenzyl isobutyrate			
Styralyl propionate	See α-methylbenzyl propionate			
Styrax, extract	[*Liquidambar* spp.]	3037	FDA §182.1	Natural flavor
Styrax, gum	See storax [*L. orientalis* Mill. or *L. styraciflua* (L.)]			
Styrene		3233		
Styryl carbinol	See cinnamyl alcohol			
2-Styryl-*m*-dioxolane	See cinnamaldehyde ethylene glycol acetal			
2-Styryl-1,3-dioxolane	See cinnamaldehyde ethylene glycol acetal			
Succinic acid	1,4-Butanedioic acid		FDA §182.1	GRAS, miscellaneous and/or general-purpose food additive

REGULATORY STATUS OF DIRECT FOOD ADDITIVES (continued)

Material	Alternatate Name [Source]	FEMA No.	Regulation	Limitations
Succinic acid, disodium salt	See disodium succinate			
Succinic anhydride	Succinic acid anhydride		FDA §121.1031	Modifier for food starch, up to 4.0%; with epichlorhydrin, up to 0.3%
Succinylated monoglycerides			FDA §121.1195	1. Emulsifier in liquid and plastic shortenings, 3% max by weight of shortening 2. Dough conditioner in bread baking, 0.5% max by weight of flour Product specifications apply
Succistearin	Stearoyl propylene glycol hydrogen succinate		FDA §121.1197	Emulsifier in or with shortenings and edible oils for use in cakes, cake mixes, fillings, icings, pastries, and toppings; product specifications apply
Sucrose	Sugar		FDA §8.303	In production of caramel
			MID	To flavor sausage, ham, and miscellaneous meat products
Sucrose acetate isobutyrate			Under FDA review	
Sucrose octaacetate		3038	FDA §172.515	Synthetic flavor
Sugar	See sucrose			
Sugar beet extract, flavor base	[*Beta vulgaris* (L.)]	—	FDA §121.1086	Flavor in food; product specifications apply
Sulfur dioxide	Sulfurous anhydride	3039	FDA §182.1	GRAS, chemical preservative; not for use in meats or in foods recognized as source of vitamin B_1
			FDA	200 to 300 ppm in molasses, dried fruits, and foods that are not good sources of vitamin B_1
			FDA §121.1031	Bleaching agent for food starch, up to 0.05%; with ammonium persulfate, up to 0.075%
Sulfuric acid	Oil of vitriol		FDA §182.1	GRAS, miscellaneous and/or general-purpose food additive

			FDA §121.1031	Modifier for food starch
			FDA §8.303	In production of caramel
			FDA §8.303	In production of caramel
Sulfurous acid	Sulfur dioxie solution			
Sulfurous anhydride	See sulfur dioxide			
Sulfurous oxide	See sulfur dioxide			
Sweet bay	See bay, sweet			
Sweet flag	See calamus [*Acorus calamus* (L.)]			
Sweet Mary	See costmary [*Chrysanthemum balsamita* (L.)]			
Sweet spirit of niter	See ethyl nitrite			
Synthetic sweet birch oil	See methyl salicylate			
Synthetic teaberry oil	See methyl salicylate			
Synthetic wintergreen oil	See methyl salicylate			
Syringa aldehyde	See *p*-tolyacetaldehyde			
Syringol	See 2,6-dimethoxyphenol			
Tagetes, meal and extract	Aztec marigold [*Tagetes erecta* (L.)]		FDA §8.306	To enhance yellow color of chicken skin and eggs, incorporated in chicken feed; ethoxyquin, 150 ppm max in finished feed; product specifications apply
Tagetes, oil	Marigold [*T. patula* (L.); *T. erecta* (L.); or *T. minuta* (L.) (*T. gladulifera* Schrank)]	3040	FDA §172.510	Flavoring use as an oil only
Tamarind	[*Tamarindus indica* (L.)]		FDA §182.1	GRAS, natural flavor extractive
Tangan-tangan, oil	See castor oil [*Ricinus communis* (L.)			
Tangerine	See mandarin			
Tangerine, oil	[*Citrus reticulata* Blanco]	3041	FDA §182.1	Natural flavor
Tannic acid	Tannin [Nutgalls of *Quercus infectoria* Oliver and related *Quercus* spp.; also in many other plants]	3042	FDA §182.1	GRAS, natural flavor extractive
Tannin	See tannic acid			
Tansy	[*Tanacetum vulgare* (L.)]	—	FDA §182.1	Natural flavor in alcoholic beverages only; thujone-free
Tansy, oil	[*T. vulgare* (L.)]	—	FDA §172.510	Natural flavor
Tar, oil	See pine tar, oil [*Pinus palustris* and other *Pinus* spp.]			
Tarragon	[*Artemisia dracunculus* (L.)]	3043		
Tarragon, oil	See estrogon, oil [*A. dracunculus* (L.)]			
Tartaric acid		3044	FDA §182.1	Natural flavor

REGULATORY STATUS OF DIRECT FOOD ADDITIVES (continued)

Material	Alternatate Name [Source]	FEMA No.	Regulation	Limitations
Tea	[*Thea sinesis* (L.)]		FDA §182.1	GRAS, natural flavor extractive
Tea tree oil		—		
1,4(8)-Terpadiene	See terpinolene			
Terpeneless Siberian pine needle oil	See bornyl acetate			
Terpene resins, natural			FDA §8.390	Diluent in inks for marking fruits and vegetables; complying with §121.1059
Terpene resins, synthetic	Polymers of α-and β-pinene		FDA §8.390	Diluent in inks for marking fruits and vegetables
Terpilene	See α-terpinene			
α-Terpilenol	See α-terpineol			
Terpinene	See terpinolene			
α-Terpinene		—	FDA §172.510	Synthetic flavor
α-Terpinene		—	FDA §172.515	Synthetic flavor
4-Terpinenol	See 4-carvomethenol			
Terpineol	See 4-carvomenthenol			
1-Terpineol	See *p*-Menth-3-en-1-ol			
α-Terpineol	*p*-Menth-1-en-8-ol	3045	FDA §172.515	Synthetic flavor
β-Terpineol	See *p*-Menth-8-en-1-ol	—	FDA §172.515	Synthetic flavor/adjuvant
Terpineol schlechtin	See α-terpineol			
α-Terpinine	See *p*-Mentha-1,4-diene		FDA §172.515	Synthetic flavor/adjuvant
γ-Terpinine	See *p*-Mentha-1,4-diene		FDA §172.515	Synthetic flavor/adjuvant
β-Terpinol	See β-terpineol			
Terpinolene	*p*-Menth-1,4(8)-diene	3046	FDA §172.515	Synthetic flavor
Terpinyl acetate	*p*-Menth-1-en-8-yl acetate	3047	FDA §172.515	Synthetic flavor
Terpinyl anthranilate	*p*-Menth-1-en-8-yl anthranilate	3048	FDA §172.515	Synthetic flavor
Terpinyl butyrate	*p*-Menth-1-en-8-yl butyrate	3049	FDA §172.515	Synthetic flavor
Terpinyl cinnamate	*p*-Menth-1-en-8-yl cinnamate	3051	FDA §172.515	Synthetic flavor
Terpinyl formate	*p*-Menth-1-en-8-yl formate	3052	FDA §172.515	Synthetic flavor
Terpinyl isobutyrate	*p*-Menth-1-en-8-yl isobutyrate	3050	FDA §172.515	Synthetic flavor
Terpinyl isopentanoate	See terpinyl isovalerate			
Terpinyl isovalerate	*p*-Menth-1-en-8-yl isovalerate	3054	FDA §172.515	Synthetic flavor
Terpinyl isovalerianate	See terpinyl isovalerate			

Terpinyl β-phenylacrylate	Seeterpinyl cinnamate			
Terpinyl 3-phenylpropenoate	See terpinyl cinnamate			
Terpinyl propionate	*p*-Menth-1-en-8-yl propionate	3053	FDA §172.515	Synthetic flavor
Terra japonica	See catechu, powder [*Acacia catechu* Willd.]		—	
Tetradecanal	See myristaldehyde			
Δ-Tetradecalactone	5-Hydroxytetradecanoic acid lactone	3590		See *Food Technol.*, February 1978
Tetradecanoic acid	See myristic acid			
Tetradecanedioc acid cyclic ethylene glycol diester	See ethylene brassylate			
Tetradecyl aldehyde	See myristaldehyde			
Tetrahydrocarvone	See *p*-menthan-2-one			
4,5,6,7-Tetrahydro-3,6-dimethylbenzofuran	Menthofuran	3235		
Tetrahydro-2-furancarbinol	See tetrahydrofurfuryl alcohol			
Tetrahydro-2-furanmethanol	See tetrahydrofurfuryl alcohol			
Tetrahydrofurfuryl acetate		3055	FDA §172.515	Synthetic flavor
Tetrahydrofurfuryl alcohol	Tetrahydro-2-furan-methanol; tetrahydro-2-furan-carbinol; THFA	3056	FDA §172.515	Synthetic flavor
Tetrahydrofurfuryl butyrate		3057	FDA §172.515	Synthetic flavor
Tetrahydrofurfuryl cinnamate	Cinnamic acid, tetrahydrofurfuryl ester	3320		See *Food Technol.*, May 1972
Tetrahydrofurfuryl 3-phenylpropenoate	See tetrahydrofurfuryl cinnamate			
Tetrahydrofurfuryl propionate		3058	FDA §172.515	Synthetic flavor
Tetrahydro-2-furylmethanol	See tetrahydrofurfuryl alcohol			
Tetrahydro-2-furylmethyl cinnamate	See tetrahydrofurfuryl cinnamate			
Tetrahydro-2-furylmethyl 3-phenylpropenoate	See tetrahydrofurfuryl cinnamate			
Tetrahydrogeraniol	See 3,7-dimethyl-1-octanol			
Tetrahydro-pseudo ionone	6,10-Dimethyl-9-undecen-2-one	3059	FDA §172.515	Synthetic flavor
Tetrahydrolinalool	3,7-Dimethyloctan-3-ol	3060	FDA §172.515	Synthetic flavor

REGULATORY STATUS OF DIRECT FOOD ADDITIVES (continued)

Material	Alternatate Name [Source]	FEMA No.	Regulation	Limitations
Tetrahydro-4-methyl-2-(2-methylpropen-1-yl) pyran	Rose oxide	3236		
Tetrahydropyrrole	See pyrrolidine			
5,6,7,8-Tetrahydroquinoxaline	Cyclohexapyrazine	3321		See *Food Technol.*, May 1972
3-Tetrahydrothiophenone	See 4,5-dihydro-3(2-*H*) thiophenone			
Tetrahydrothiophen-3-one	See 4,5-dihydro-3(2*H*) thiophenone			
α,2,6,6-Tetramethyl-1,3-cyclohexadien-1-propanol	See dehydrodihydroionol			
4-(2,5,6,6-Tetramethyl-2-cyclohexen-1-yl)-3-buten-2-one	See α-irone			
Tetramethyl ethylcyclohexenone	Mixture of 5-ethyl-2,3,4,5-tetramethyl-2-cyclohexen-1-one and 5-ethyl-3,4,5,6-tetramethyl-2-cyclohexen-1-one	3061	FDA §172.515	Synthetic flavor/adjuvant
1,5,5,9-Tetramethyl-13-oxatricyclo(8.3.0.0(4,9) tridecane		3471		
2,3,5,6-Tetramethylpyrazine		3237		10 ppm max in finished food
Tetrasodium pyrophosphate			FDA §19	Emulsifier, up to 3% in specified cheeses
3,3′-Tetrathio-bis(2-methylfuran)	See bis(2-methyl-3-furyl) tetrasulfide			
Tetryl formate	See isobutyl formate			
THBP	2,4,5-Trihydroxybutyrophenone		FDA §121.2001	Antioxidant in manufacture of food-packaging material; 0.005% max addition to food
			FDA §121.116	Antioxidant, alone or with other permitted antioxidants; total antioxidant(s), 0.02% max of oil or fat content of the food, including essential

				(volatile) oil content
Thein	See caffeine			
Theobromine	3,7-Dihydro-3,7-dimethyl-1H-purine-2,6-*dione*	3591		See *Food Technol.*, February 1978
	3,7-Dimethylxanthine	3591		See *Food Technol.*, February 1978
THFA	See tetrahydrofurfuryl alcohol			
Thiamine hydrochloride	Vitamin B_1; vitamin B_1 hydrochloride	3322	FDA §182.1	GRAS, nutrient and/or dietary supplement For FEMA max use levels, see *Food Technol.*, May 1972
Thiamine mononitrate	Vitamin B_1 mononitrate		FDA §182.1	GRAS, nutrient and/or dietary supplement
2-Thiapropane	See methyl sulfide			
Thibetolide	See ω-pentadecalactone			
2-Thienyl disulfide	2,2-Dithiodithiophene	3323		See *Food Technol.*, May 1972
2-Thienyl mercaptan	2-Thienylthiol	3062	FDA §172.515	Synthetic flavor
Thienyl pyrimidine	See 5,7-dihydro-2-methylthieno (3,4-D)pyrimidine			
2-Thienylthiol	See 2-thienyl mercaptan			
Thioacetic acid, ethyl ester	See ethyl thioacetate			
Thioallyl ether	See allyl sulfide			
Thiobenzyl alcohol	See benzyl mercaptan			
Thiobutyric acid, methyl ester	See methyl thiobutyrate			
2,2′-(Thiodimethylene)-difuran	2-Furfuryl monosulfide	3238		
Thiodipropionic acid			FDA §121.2001	Antioxidant in manufacture of food-packaging materials; 0.005% max addition to food
			FDA §182.1	GRAS, chemical preservative; total antioxidants, 0.02% max of fat or oil content, including essential (volatile) oil content of food
Thiofuroic acid, methyl ester	See methyl thiofuroate			
Thiogeraniol		3472		
Thioguaiacol	See *o*-methylthio)-phenol			
Thiolactic acid	See 2-mercaptopropionic acid			

REGULATORY STATUS OF DIRECT FOOD ADDITIVES (continued)

Material	Alternatate Name [Source]	FEMA No.	Regulation	Limitations
2-Thiolpropionic acid	See 2-mercaptopropionic acid			
Thiomethyl alcohol	See methyl mercaptan			
2-Thionaphthol	See 2-naphthalenthiol			
Thiophenol	See benzenethiol			
3-Thiophenone	See 4,5-dihydro-3(2*H*) thiophenone			
n-Thiopropyl alcohol	See propyl mercaptan			
Thistle, blessed	[*Cnicus benedictus* (L.)]	—	FDA §172.510	Natural flavor
Thistle, blessed, extract	[*C. benedictus* (L.)]	—	FDA §172.510	
Thistle, blessed, extract solid	[*C. benedictus* (L.)]	—	FDA §172.510	
Thistle, blessed, oil	[*C. benedictus* (L.)]	—	FDA §172.510	
Threonine (L-and DL-forms)			FDA §182.1	GRAS, nutrient and/or dietary supplement, as free acid, hydrochloride salt, hydrate, or anhydrous form
4-Thujanol	Sabinenehydrate	3239		
Thyme, oil	[*Thymus vulgaris* (L.)]	3064	FDA §182.1	Natural flavor
Thyme, white	[*T. vulgaris* (L.); *T. zygis* (L.) var. *gracilis* Boiss.]	3063	FDA §182.1	GRAS, natural flavor extractive
Thyme, white, oil	[*T. vulgaris* (L.)]	3065	FDA §182.1	Natural flavor
Thyme, wild or creeping	[*T. serpyllum* (L.)]		FDA §182.1	GRAS, natural flavor and natural flavor extractive
Thymol	3-*p*-Cymenol	3066	FDA §172.515	Synthetic flavor
Thymol (iso)	See carvacrol			
Thymol methylether	See 1-methyl-3-methoxy-4-isopropylbenzene			
Thymus capitatus Hoff. and Link	Spanish "origanum"		FDA §172.510	Natural flavor
Tiglaldehyde	See 2-methyl-2-butenal			
Tirpilene	See *p*-mentha-1,3-diene			
Titanium dioxide	Unitane		FDA §8.316	1. Coloring foods generally, except standardized foods unless the standards authorize its use; 1% max 2. Mixtures containing titanium dioxide

				may contain diluents listed in regulations for color additives for food use, silicon dioxide and/or aluminum oxide as dispersing aids (SiO_2 and Al_2O_3 at 2% max)
				Product specifications apply
			FDA §8.6005	Coloring for ingested and externally applied drugs generally; external application includes use in area of the eye; specifications and diluents as in §8.316
Toasted, partially defatted, cooked cottonseed flour	See cottonseed flour, toasted, partially defatted, cooked			
α-Tocopherol acetate			FDA §182.1	GRAS, nutrient and/or dietary supplement
Tocopherols			FDA	0.03% max of fat content of food
			FDA §182.1	GRAS, nutrient and/or dietary supplements and chemical preservatives
			MID	To retard rancidity in rendered animal fat or in combination of such fat and vegetable fat, up to 0.03%, 30% concentration of tocopherols in vegetable oils shall be used when added as antioxidant to products designated as lard or rendered pork fat
α-Tolualdehyde	See phenylacetaldehyde			
Tolualdehyde glyceryl acetal, mixed *o, m, p*		3067	FDA §172.515	Synthetic flavor
Tolualdehydes, mixed *o, m, p*		3068	FDA §172.515	Synthetic flavor
Tolu balsam, extract	[*Myroxylon balsamum* (L.) Harms; *M. toluiferum* H.B.K.]	3069	FDA §172.510	Natural flavor
Tolu balsam, gum	[*Myroxylon* spp.]	3070	FDA §172.510	Natural flavor
o-Toluenethiol	2-Methylthiophenol	3240		
α-Toluenethiol	See benzyl mercaptan			
α-Toluic acid	See phenylacetic acid			
α-Toluic acid, ethyl ester	See ethyl phenylacetate			
Toluic aldehyde, mixed 2,3,4	See tolualdehydes, mixed *o, m, p*			

REGULATORY STATUS OF DIRECT FOOD ADDITIVES (continued)

Material	Alternatate Name [Source]	FEMA No.	Regulation	Limitations
α-Toluic aldehyde	See phenylacetaldehyde			
p-Tolylacetaldehyde	*p*-Methylphenylacetaldehyde	3071	FDA §172.515	Synthetic flavor
o-Tolyl acetate	*o*-Cresyl acetate	3072	FDA §172.515	Synthetic flavor
p-Tolyl acetate	*p*-Cresyl acetate	3073	FDA §172.515	Synthetic flavor
4-(*p*-Tolyl)-2-butanone	*p*-Methylbenzylacetone	3074	FDA §172.515	Synthetic flavor
p-Tolyl dodecanoate	See *p*-tolyl laurate			
p-Tolyl ethanoate	See *p*-tolyl acetate			
1-*p*-Toly-1-ethanol	See *p*,α-dimethylbenzyl alcohol			
o-Tolyl isobutyrate		—		
p-Tolyl isobutyrate	*p*-Cresyl isobutyrate	3075	FDA §172.515	Synthetic flavor
p-Tolyl isovalerate	See *p*-tolyl 3-methylbutyrate			
p-Tolyl laurate	*p*-Tolyl dodecanoate	3076	FDA §172.515	Synthetic flavor
o-Tolylmercaptan	See *o*-toluenethiol			
p-Tolyl 3-methylbutyrate		3387		
p-Tolyl methyl carbinol	See *p*,α-dimethylbenzyl alcohol			
p-Tolyl methyl ketone	See 4′-methylacetophenone			
o-Tolyl 2-methylpropanoate	See *o*-tolyl isobutyrate			
p-Tolyl 2-methylpropanoate	See *p*-tolyl isobutyrate			
p-Tolyl octanoate			Under FEMA review	
p-Tolyl phenylacetate	*p*-Cresyl phenylacetate	3077	FDA §172.515	Synthetic flavor
2-*p*-Tolyl-2-propanol	See *p*,α,α-trimethylbenzyl alcohol			
3-*p*-Tolylpropenal	See *p*-methyl cinnamaldehyde			
2-*p*-Tolylpropene	See *p*,α-dimethylstyrene			
2-(*p*-Tolyl)propionaldehyde	*p*-Methylhydratropicaldehyde	3078	FDA §172.515	Synthetic flavor
p-Tolyl α-toluate	See *p*-tolyl phenylacetate			
Tonkalide	See γ-hexalactone			
Tonquin musk	See musk			
Tragacanth, gum	[*Astragalus gummifier* Lab.; or other Asiatic *Astragalus* spp.]	3079	FDA §182.1	GRAS, stabilizer
Tragacanth, Indian	See karaya, gum [*Sterculia urens* Roxb.]			

Triacetin	See (tri-)acetin			
Tributyl acetylcitrate		3080	FDA §172.515	Synthetic flavor
Tributyrin	See (tri-)butyrin			
Tricalcium phosphate		3081	FDA §22.8 and §22.9	In vanilla powder and vanilla-vanillin powder
			FDA §15	Bleaching agent in flour, at not more than 6 parts by weight, alone or in combination with potassium alum, calcium sulfate, magnesium carbonate, sodium aluminum sulfate, dicalcium phosphate, starch, and/or calcium carbonate, and with 1 part by weight of benzoyl peroxide
			FDA §182.1	GRAS, dietary supplement
			MID	Aid in rendering animal fats
Tricalcium silicate			FDA §182.1	GRAS, anticaking agent in table salt up to 2%
Trichloroethane	See trichloroethylene			
Trichloroethylene	Trichloroethane		FDA §8.305	Solvent for annatto; residues, 30 ppm max; total residues of chlorinated solvents, 30 ppm max
Trichloromethane	See chloroform			
Tridecanedioic acid cyclic ethylene glycol diester	See ethyl brassylate			
2-Tridecanone		3388		
2-Tridecenal		3082	FDA §172.515	Synthetic flavor
Triethyl citrate	Ethyl citrate	3083	FDA §182.1	GRAS, miscellaneous and/or general-purpose food additive in dried egg whites, up to 0.25%
Triethyl 2-hydroxy-1,2,3-propanetricarboxylate	See triethyl citrate			
Trifolium, extract solid	See clover, tops, red, extract solid [*Trifolium pratense* (L.) and other *Trifolium* spp.]			
Trihydroxypropane	See glycerol			
Trimethylamine		3241		
p-α,α-Trimethylbenzyl alcohol	*p*-Cymen-8-ol; dimethyl-*p*-tolyl carbinol	3242		

REGULATORY STATUS OF DIRECT FOOD ADDITIVES (continued)

Material	Alternate Name [Source]	FEMA No.	Regulation	Limitations
1,3,3-Trimethyl bicyclo (2.2.1)heptan-2-ol	See fenchyl alcohol			
1,7,7-Trimethylbicyclo (2.2.1)heptan-2-2ol	See borncol			
1,7,7-Trimethylbicyclo (2.2.1) heptanone	See *d*-camphor			
1,3,3-Trimethylbicyclo (2.2.1) heptan-2-one	See *d*-fenchone			
2,6,6-Trimethylbicyclo-(3,1,1)-2-heptene	See α-pinene			
2,2,6-Trimethyl-1,3-cyclohexadien-1-carboxaldehyde	See 2,6,6-trimethyl cyclohexa-1,3-dienyl methanal			
2,2,6-Trimethyl-4,6-cyclohexadien-1-carboxaldehyde	See 2,6,6-trimethyl cyclohexa-1,3-dienyl methanal			
4-(2,6,6-Trimethylcyclohexadien-1-yl)-2-butanone	See dehydrodihydroionone			
4-(2,6,6-Trimethylcyclohexa-1,3-dienyl)-but-2-en-4-one		3420		
2,6,6-Trimethylcyclohexa-1,3-dienyl methanal		3389		
3,5,5-Trimethyl-1,w-cyclohexanedione	See 2-hydroxy-3,5,5-trimethyl-2-cyclohexenone			
2,2,6-Trimethylcyclohexanone		3473		
2,6,6-Trimethyl-1-cyclohexen-1-acetaldehyde		3474		
3,5,5-Trimethyl-2-cyclohexen-1-one	See isophorone			
2,6,6-Trimethylcyclohex-2-ene-1,4-dione		3421		
1-(2,6,6-Trimethyl-2-cyclohexene-1-yl)-1-6-hepta-	See allyl α-ionone			

dien-3-one			
3,5,5-Trimethyl-2-cyclohexen-1-one	See isophorone		
4-(2,6,6-Trimethyl-1-cyclohexen-1-yl)-3-buten-2-one	See β-ionone		
4-(2,6,6-Trimethyl-2-cyclohexene-1-yl)-3-buten-2-one	See α-ionone		
4-(2,6,6)-Trimethylcyclohex-1-enylbut-2-en-4-one		3243	
1-(2,6,6-Trimethyl-3-cyclohexen-1-yl)-1,6-heptadien-3-one	See allyl α-ionone		
4-(2,6,6-Trimethyl-2-cyclohexen-1-yl)-3-methyl-3-buten-2-one	See α-somethylionone		
5-(2,6,6-Trimethyl-1-cyclohexen-1-yl)-4-penten-3-one	See methyl β-ionone		
5-(2,6,6-Trimethyl-2-cyclohexen-1-yl)-4-penten-3-one	See methyl α-ionone		
5-(2,6,6-Trimethyl-3-cyclohexen-1-yl)-4-penten-3-one	See methyl δ-ionone		
2,3,3-Trimethylcyclopent-3-en-1-ylactaldehyde	*Campholenic* aldehyde	3592	See *Food Technol.*, February 1978
2,4,5-Trimethyl-2,5-dihydrooxazole	See 2,4,5-trimethyl-δ-3-oxazoline		
2,2,4-Trimethyl-1,3-dioxolane	See 2,2,4-trimethyl-1,3-oxacyclopentane		
1,3,7-Trimethyl-2,6-dioxopurine	See caffeine		
3,7,11-Trimethyl-1,6,10-dodecatrien-3-ol	See nerolidol		

REGULATORY STATUS OF DIRECT FOOD ADDITIVES (continued)

Material	Alternate Name [Source]	FEMA No.	Regulation	Limitations
3,7,11-Trimethyl-2,6,10-dodecatrien-1-ol	See farnesol			
1,1,3-Trimethyl-2-formyl-cyclohexa-2,4-diene	See 2,6,6-trimethylcyclohexa-1,3-dienyl methanal			
3,5,5-Trimethylhexanal		3524		
3,5,5-Trimethyl-hexanol		3324		See *Food Technol.*, May 1972
Trimethylhexyl alcohol	See 3,5,5-trimethyl-1-hexanol			
4,12,12-Trimethyl-9-methylene-5-oxatricyclo(8.2.0.0(4,6)) dodecane	See β-caryophyllene oxide			
1,3,3-Trimethyl-2-norbornanol	See fenchyl alcohol			
d-1,3,3-Trimethyl-2-norbornanone	See *d*-fenchone			
1,3,3-Trimethyl-2-norbornanyl acetate		3390		
d-1,3,3-Trimethyl-2-norcamphanone	See *d*-fenchone			
2,2,4-Trimethyl-1,3-oxacyclopentane		3441		
2,4,5-Trimethyl-δ-3-oxazoline		3525		
2,6,10-Trimethyl-2,6,10-pentadecatrien-14-one		3442		
6,10,14-Trimethyl-5,9,13-pentadecatrien-2-one	See 2,6,10-trimethyl-2,6,10-pentadecatrien-14-one			
2,3,5-Trimethylpyrazine		3244		10 ppm max in finished food
2,4,5-Trimethylthiazole		3325		see *Food Technol.*, May 1972
1,3,7-Tripropylxanthine	See caffeine			
Tripropionin	See glyceryl tripropanoate			
Trisodium citrate	See sodium citrate			
Trisodium phosphate			FDA §19	Emulsifier, up to 3% by weight of spec-

				ified cheeses
			MID	In rendering animal fats
5-Trithiane 2,2,4-4,6,6-hexamethyl	See trithioacetone			
Trithioacetone		3475		
Triticum	See dog grass, extract [*Agropyron repens* (L.) Beauv.]			
Tryptophan (L- and DL-forms)			FDA §182.1	GRAS, nutrient and/or dietary supplement, as free acid, hydrochloride salt, hydrate, or anhydrous form
Tuberose, oil	[*Polianthes tuberosa* (L.)]	3084	FDA §182.1	GRAS, natural flavor extractive
Tuberyl acetate	See dihydrocarvyl acetate			
Tunu	Tuno [*Castilla fallax* Cook]		FDA §121.1059	Natural (coagulated or concentrated latex) masticatory substance in chewing-gum base
Turmeric	Tumeric [*Curcuma longa* (L.)]	3085	FDA §182.1	GRAS, natural flavor and natural flavor extractive
			FDA §8.309	Food use generally, except in standardized foods where standards do not authorize its use
			MID	To color sausage casings, oleomargarine, or shortening, and in branding or marking ink for products; may be mixed with approved synthetic dyes or harmless inert materials, such as common salt and sugar
Turmeric, extract	[*Curcuma longa* (L.)]	3086	FDA §182.1	Natural flavor
Turmeric, oleoresin	[*Curcuma longa* (L.)]	3087	FDA §8.310	Food use generally, except in standardized foods where standards do not authorize its use; residues of solvents are limited by applicable food additive orders; product specifications apply
			FDA §182.1	Natural flavor
Turpentine, gum	[*Pinus* spp.]	3088	FDA §172.510	
Turpentine, steam distilled	[*Pinus palustris* Mill. and other *Pinus* spp. that yield terpene oils exclusively]	3089	FDA §172.510	Natural flavoring substances and natural substances in conjunction with flavors
Tween® 20	See polysorbate 20			

REGULATORY STATUS OF DIRECT FOOD ADDITIVES (continued)

Material	Alternate Name [Source]	FEMA No.	Regulation	Limitations
Tween 60	See polysorbate 60			
Tween 80	See polysorbate 80			
Tyrosine (L- and DL-forms)			FDA §182.1	GRAS, nutrient and/or dietary supplement as free acid, hydrochloride salt, hydrate, or anhydrous form
Ultramarine blue			FDA §8.319	In coloring salt intended for animal feed, 0.5% max by weight of salt; product specifications apply
Ultraviolet radiation for processing and treatment of food			FDA §121.3006	1. Control of surface microorganisms on food and food products 2. Sterilization of potable water used in food production Product specifications apply
2,4-Undecadienal		3422		
2,5-Undecadienal		—		
2,3-Undecadione	Acetyl nonyryl	3090	FDA §172.515	Synthetic flavor
γ-Undecalactone	4-Hydroxyundecanoic acid, γ-lactone; peach aldehyde; aldehyde C-14 (so-called)	3091	FDA §172.515	Synthetic flavor
δ-Undecalactone	See 5-hydroxyundecanoic acid lactone			
Undecanal	Aldehyde C-11 undecylic	3092	FDA §172.515	Synthetic flavor
Undecanoic acid		3245		
1-Undecanol	See undecyl alcohol			
2-Undecanol		3246		
Undecanolide-1,4	See γ-undecalactone			
2-Undecanone	Methyl nonyl ketone	3093	FDA §172.515	Synthetic flavor
2-Undecenal		3423		
9-Undecenal	Undecenoic aldehyde	3094	FDA §172.515	Synthetic flavor
10-Undecenal	Aldehyde C-11 undecylenic; hendecenal; undecylenic aldehyde	3095	FDA §172.515	Synthetic flavor
2-Undecen-1-al	See 2-undecenal			
10-Undecenoic acid		3247		

Undecenoic aldehyde	See 9-undecenal			
Undecen-1-ol	Undecylenic alcohol	—	FDA §172.515	Synthetic flavor/adjuvant
2-Undecen-1-ol		—		
Undecenyl acetate	See 10-undecen-1-yl acetate			
10-Undecen-1-yl acetate	10-Hendecenyl acetate; undecenyl acetate	3096	FDA §172.515	Synthetic flavor
n-Undecoic acid	See undecanoic acid			
Undecyl alcohol	1-Undecanol	3097	FDA §172.515	Synthetic flavor
n-Undecylaldehyde	See undecanal			
Undecyl carbinol	See lauryl alcohol			
Undecylenic acetate	See 10-undecen-1-yl acetate			
Undecylenic acid	See 10-undecenoic acid			
Undecylenic alcohol	See undecen-1-ol			
Undecylenic aldehyde	See 9-undecenal			
Undecylenic aldehyde	See 10-undecenal			
n-Undecylic acid	See undecanoic acid			
Sec-Undecylic alcohol	See 2-undecanol			
Undecylic aldehyde	See undecanal			
γ-Undecyl lactone	See γ-undecalactone			
Valencene		3443		
Valeral	See valeraldehyde			
Valeraldehyde	Pentanal	3098	FDA §172.515	Synthetic flavor
Valeraldehyde (iso)	See 3-methylbutyraldehyde			
Valerianic acid	See valeric acid			
Valerianic acid	See isovaleric acid			
Valerian, rhizome and root, extract	[*Valeriana officinalis* (L.)]	3099	FDA §172.510	Natural flavoring substances and natural substances used in conjunction with flavor
Valerian, *root, oil*	[*V. officinalis* (L.)]	3100	FDA §172.510	
Valeric acid	Pentanoic acid	3101	FDA §172.515	Synthetic flavor
Valeric (iso) acid	See isovaleric acid			
Valeric aldehyde	See valeraldehyde			
Valeric (iso) aldehyde	See 3-methylbutyraldehyde			
3-Valerolactone	See γ-valerolactone			
4-Valerolactone	See γ-valerolactone			
γ-Valerolactone	4-Hydroxypentanoic acid	3103		
Valeryl acetal	See 2,3-heptanedione			
γ-Valeryllactone	See γ-valerolactone			

REGULATORY STATUS OF DIRECT FOOD ADDITIVES (continued)

Material	Alternate Name [Source]	FEMA No.	Regulation	Limitations
Validol	See menthyl isovalerate			
Valine (L- and DL-forms)		3444	FDA §182.1	GRAS, nutrient and/or dietary supplement, as free acid, hydrochloride salt, hydrate, or anhydrous form
Vanay	See (tri-)acetin			
Vanilla	[*Vanilla planifolia* Andr.; *V. tahitensis* J. W. Moore]	3104	FDA §182.1	GRAS, natural flavoring and natural flavor extractive
Vanilla, extract	[*Vanilla* spp.]	3105	FDA §182.1	Natural flavor
Vanillaldehyde	See vanillin			
Vanilla, oleoresin	[*Vanilla* spp.]	3106	FDA §182.1	Naturalflavor
Vanillic aldehyde	See vanillin			
Vanillin	Methylprotocatechuic aldehyde	3107	FDA §182.1	GRAS, synthetic flavor and adjuvant
Vanillin acetate	Acetyl vanillin	3108	FDA §172.515	Synthetic flavor
Vanillin methyl ether	See veratraldehyde			
Vanillylacetone	See zingerone			
Vanillyl alcohol		—		
Vanillylidene acetone		—		
Vegetable juice			FDA §8.314	Food use generally, except for standardized foods where standards do not authorize its use; product specifications apply
Vegetable oils or fats			FDA §8.305	Extractants for annatto
Veltol®-Plus	See ethyl maltol			
Venezuelan chicle	[*Manilkara williamsii* Standley and related spp.]		FDA §121.105	Natural (coagulated or concentrated latex) masticatory substance in chewing-gum base
Veratraldehyde	Vanillin methyl ether	3109	FDA §172.515	Synthetic flavor
Veratric aldehyde	See veratraldehyde			
Verbenol	4-Hydroxy-2,6,6-trimethylbicyclo[3.1.1]hept-2-ene	3594	FDA §172.515	Synthetic flavor; see *Food Technol.*, February 1978
	2-Pinen-4-ol	3594		See *Food Technol.*, February 1978
Veronica	[*Veronica officinalis* (L.)]	—	FDA §172.510	Natural flavor in alcoholic beverages only

Vervain, European	[*Verbena officinalis* (L.)]	—	FDA §172.510	Natural flavor in alcoholic beverages only
Vetiver	[*Vetiveria zizanioides* Stapf.]	—	FDA §172.510	Natural flavor
Vetiver, oil	[*Vetiveria zizanioides* Stapf.]	—	FDA §172.510	
Vetiveryl acetate		—		
Vianol®	See butylated hydroxytoluene			
Vinegar naphtha	See ethyl acetate			
Vinegar naptha	See ethyl (*p*-tolyloxy)acetate			
Vinyl acetate			FDA §121.1031	Modifier for food starch; acetyl groups in food starch modified, 2.5% max
Vinyl amyl ketone	See 1-octen-3-one			
o-Vinylanisole		3248		
Vinylbenzene	See styrene			
Vinylbenzol	See styrene			
Vinyl carbinyl cinnamate	See allyl cinnamate			
p-Vinylcatechol-*O*-methyl ether	See 2-methoxy-4-vinylphenol			
Vinyl ethyl carbinol	See 1-penten-3-ol			
p-Vinylguaiacol	See 2-methoxy-4-vinylphenol			
β-Vinylhydrocinnamaldehyde	See 3-phenyl-4-pentenal			
cis- and *trans*-2-Vinyl-2-methyl-5-(1′-hydroxy-1′-methylethyl)tetrahydrofuran	See linalool oxide			
Violet, flowers and leaves	[*Viola odorata* (L.)]		FDA §182.1	GRAS, natural flavor extractive
Violet leaves, absolute	[*V. odorata* (L.)]	3110	FDA §182.1	GRAS, natural flavor extractive
Violet, Swiss	[*V. calcarata* (L.)]	—	FDA §172.510	Natural flavor in alcoholic beverages only
Virginia snakeroot	See serpentaria [*Aristolochia serpentaria* (L.)]			
Viridine	See phenylacetaldehyde dimethyl acetal			
Vitamin A	Retinol		FDA §182.1	GRAS, nutrient and/or dietary supplement
Vitamin A acetate			FDA §182.1	GRAS, nutrient and/or dietary supplement

REGULATORY STATUS OF DIRECT FOOD ADDITIVES (continued)

Material	Alternate Name [Source]	FEMA No.	Regulation	Limitations
Vitamin A palmitate			FDA §182.1	GRAS, nutrient and/or dietary supplement
Vitamin B_1	See thiamine hydrochloride			
Vitamin B_{12}			FDA §182.1	GRAS, nutrient and/or dietary supplement
Vitamin C	See ascorbic acid			
Vitamin D	Vitamin D_2, vitamin D_3		FDA §182.1	GRAS, nutrient and/or dietary supplement
Vitamin U	See *dl*-(3-amino-3-carboxypropyl) dimethylsulfonium chloride			
Volatile solvents			FDA §8.201	Diluent in aqueous suspension of Citrus Red No. 2; no residue after application to oranges
Walnut hull, extract	[*Juglans nigra* (L.); *J. regia* (L.)]	3111	FDA §172.510	
Walnut husks (hulls), leaves, and green nuts	[*J. nigra* (L.); *J. regia* (L.)]		FDA §172.510	Natural flavor
Walnut leaves, extract	[*J. nigra* (L.); *J. regia* (L.)]	—	FDA §172.510	
West Indian bay	See bay leaf, West Indian			
West Indian sandalwood	Amyris [*Amyris balsamifera* (L.)]		FDA §172.510	Natural flavor
Whey, dried			MID	To bind and extend imitation sausage, nonspecific loaves, soups, and stews
White flag, extract	See orris root, extract [*Iris germanica* (L.), including its variety *Florentina* Dykes]			
White mineral oil			FDA §121.1146	1. 0.6% max in capsules or tablets containing concentrates of flavoring, spices, condiments, and nutrients intended for addition to food, excluding confectionery; release agent, binder, and lubricant 2. 0.6% max in capsules or tablets containing foods for special dietary use; release agent, binder, and lubricant

				3. Float on fermentation fluids in manufacture of vinegar and wine to prevent or retard access of air, evaporation, and wild yeast contamination during fermentation 4. Defoamer in food, in accordance with §121.1099 5. 0.15% max in bakery products; release agent and lubricant 6. 0.2% max in dehydrated fruits and vegetables; release agent 7. 0.10% max in egg white solids; release agent 8. Protective coating on raw fruits and vegetables 9. 0.095% max in frozen meats; as component of hot-melt coating 10. Protective float on brine in curing of pickles 11. 0.3% max in molding starch used in manufacture of confectionery
White vitriol	See zinc sulfate			
White wax	See beeswax, white [*Apis mellifera* (L.)]			
Whole fish protein concentrate			FDA §121.1202	In the household as protein supplement in food, in consumer-sized units not exceeding 1 lb net weight; when consumed regularly by children up to 8 years old, the additive in their total diet should not exceed 20 g/day; product specifications apply
Wild cherry, bark	See cherry bark, wild, extract [*Prunus serotina* Ehrh.]			
Wild ginger, Canadian, oil	See snakeroot, Canadian, oil [*Asarum canadense* (L.)]			
Wine			FDA-GRAS	
Wintergreen, extract	[*Gaultheria procumbens* (L.)]	3112		
Wintergreen, oil	[*G. procumbens* (L.)]	3113		

REGULATORY STATUS OF DIRECT FOOD ADDITIVES (continued)

Material	Alternate Name [Source]	FEMA No.	Regulation	Limitations
Woodruff, sweet	[*Asperula odorata* (L.)]	—	FDA §172.510	Natural flavor in alcoholic beverages only
Wormwood	See artemisia [*Artemisia absinthium* (L.) and other *Artemisia* spp.]			
Wormwood, extract	Absinthium[*Artemisia absinthium* (L.)]	3115	FDA §172.510	
Wormwood, oil	Absinthium [*Artemisia absinthium* (L.)]	3116	FDA §172.510	
Xanthan gum	[*Xanthomonas campestris*]		FDA §121.1224	Stabilizer, emulsifier, thickener, suspending agent, bodying agent, or foam enhancer in foods, provided that food standards do not preclude such use; product specifications apply
Xanthophyll	See corn endosperm oil			
2,5-Xylenol	2,5-Dimethylphenol	3595		See *Food Technol.*, February 1978
	1,Hydroxy-2,5-dimethyl-benzene	3595		See *Food Technol.*, February 1978
2,6-Xylenol	2,6-Dimethylphenol	3249		
3,4-Xylenol	3,4-Dimethylphenol	3596		See *Food Technol.*, February 1978
	1-Hydroxy-3,4-dimethyl-benzene	3596		See *Food Technol.*, February 1978
Xylitol			FDA §121.1114	In foods for special dietary uses
2,4-Xylylaldehyde	See 2,4-dimethylbenzaldehyde			
Yara yara	See β-naphthyl methyl ether			
Yarrow, herb	[*Achillea millefolium* (L.)]	3117	FDA §172.510	Natural flavor in Beverages Only: Thujone-Free
Yeasts, dried	See dried yeasts [*Saccharomyces* spp.]			
Yeast extract, autolyzed			MID	To flavor various meat products
Yellow prussiate of soda	Sodium ferrocyanide decahydrate, $Na_4Fe(CN)_6 \cdot 10H_2O$		FDA §121.1032	Anticaking agent, 5 ppm max calculated as anhydrous sodium ferrocyanide, in salt; 13 ppm max in "fine salt;" product specifications apply
Yellow wax	See beeswax, yellow			
Yerba santa	[*Eriodictyon californicum* (Hook. and Arn.) Torr.]		FDA §172.510	Natural flavor

Yerba santa, fluid extract	[*E. californicum* (Hook. and Arn.) Torr.]	3118	FDA §172.510	
Ylang-ylang, oil	[*Cananga odorata* Hook. f. and Thoms.]	3119	FDA §182.1	GRAS, natural flavor extractive
Yucca	Joshua tree [*Yucca brevifolia* Engelm.]	3120	FDA §172.510	Natural flavor
Yucca, Mohave, extract	[*Yucca schidigera* Roezl ex Ortiges (*Y. mohavensis* Sarg.)]	3121	FDA §172.510	Natural flavor
Zedoary	[*Curcuma zedoaria* (Berg.) Rosc.]	3122	FDA §182.1	GRAS, natural flavor
Zedoary, bark, extract	[*C. zedoaria* (Berg.) Rosc.]	3123	FDA §182.1	GRAS, natural flavor extractive
Zibeth	See civet			
Zinc acetate			FDA	GRAS, 1 mg max in daily dose of dietary supplements
Zinc chloride	Butter of zinc		FDA §182.1	GRAS, nutrient and/or dietary supplement
Zinc gluconate			FDA §182.1	GRAS, nutrient and/or dietary supplement
Zinc oxide			FDA §182.1	GRAS, nutrient and/or dietary supplement
Zinc stearate			FDA §182.1	GRAS, nutrient and/or dietary supplement, when prepared from stearic acid free from chick edema factor
Zinc sulfate	White vitriol		FDA §182.1	GRAS, nutrient and/or dietary supplement
Zingerone	4-(4-Hydroxy-3-methoxyphenyl)-2-butanone; 2-(4-Hydroxy-3-methoxyphenyl) ethyl methyl ketone	3124	FDA §172.515	Synthetic flavor